Springer Tracts in Natural Philosophy

Volume 20

Edelen · Wilson

Relativity and the Question of Discretization in Astronomy

with 34 Figures

Springer-Verlag New York Heidelberg Berlin 1970

Dominic G. B. Edelen

Center for the Application of Mathematics
Lehigh University, Bethlehem, USA

Albert G. Wilson

Department of Astronomy,
University of Southern California, Los Angeles, USA;
Sometime Director of the Lowell Observatory and
Staff Member of the Mt. Wilson and Palomar Observatories

Typsetting and printing: Werk- und Feindruckerei Dr. Alexander Krebs, Weinheim und Hemsbach/Bergstr.
Binding: Konrad Triltsch, Graphischer Betrieb, Würzburg

Preface

Theoretical researches in general relativity and observational data from galactic astronomy combine in this volume in contributions to one of the oldest questions of natural philosophy: Is the structure of the physical world more adequately described by a continuous or a discrete mode of representation? Since the days of the Pythagoreans, this question has surfaced from time to time in various guises in science as well as in philosophy. One of the most bitterly contested and illuminating controversies between the continuous and the discrete viewpoints is to be found in the wave versus corpuscular description of optical phenomenae. This controversy was not resolved to the satisfaction of most of its protaganists until the development of the quantum theory. However, several obscurities that still becloud the question suggest that some deeper formulation may be necessary before more satisfactory answers can be given[1]. The firm establishment of the validity of quantized structure and discrete energy distributions on the atomic scale following the ideas of Max Planck, together with the apparent absence of quantization effect in astronomical and cosmic structures leaves uncertainties concerning the role played by the scale of the observer in perceiving or not perceiving discrete distributions. Some of the metaphysical interpretations and implications of the quantum mechanics that have been made in recent years[2] would be subject to revision if the existence of discretized descriptions were to be established in astronomical and cosmic structures. Accordingly it becomes important to investigate – observationally and theoretically – the possibility of *quantization* or *discreteness* on all scales.

One of the first physicists to be concerned with large scale discretization was Sir James Jeans who pointed to the existence of quantization in *stellar sizes*[3]. Jeans held that the sizes of stars must reflect the quantized sizes of the atoms out of which they were made, "The radii of atoms

[1] See for example, the account of a colloquium on the quantum theory held at Cambridge in July 1968 in *Theoria to Theory* 3, 69 Jan. (1969).

[2] Mario Bunge (ed.): *Quantum Theory and Reality*. New York: Springer-Verlag 1967.

[3] Jeans, J. H.: *Astronomy and Cosmogony*. Cambridge University Press 1928.

which have 0, 1, 2, 3, ... rings of electrons revolving round their nuclei are in the proportion $0^2 : 1^2 : 2^2 : 3^2 : ..$, and if all the stars of a given mass are graded according to size, we find that they fall into groups in which the radii are in someting like these same proportions." Jeans goes on to say, "... theory and observations both show that the groups of configurations corresponding to the different atomic radii remain distinct only in unusually massive stars. In stars of moderate mass the distinction becomes blurred, so that the various types of configuration merge continuously into one another, but we have found that the diameters of stars of large mass reflect quite clearly the different possible diameters of the atoms of which they are composed". Except for the discrimination between giants and dwarfs, this statement by Jeans has never been definitively confirmed or refuted. Olin Eggen in a series of studies of magnitude-color diagrams of open star clusters[4] reported that he had detected several parallel sub-sequences of stars associated with the main sequence. Eggen's results were never definitively established and it was later pointed out[5] that if corrections were made for spectral line blanketing effects, the various subdwarf sequences disappeared.

The possibility of a discrete distribution in the masses of galaxies has long been recognized. There exist at least two peaks in the frequency distribution of masses of galaxies, one at around 10^{11} solar masses and one at around 10^9 solar masses[6]. In addition, evidence that spherical (*EO*) galaxies possessed a discrete distribution in size was reported in 1950 by Wilson[7], "From a preliminary study made of the sizes of globular nebulae in the Coma, Ursa Major I, Corona, and Ursa Major II clusters, sizes being taken both from tracings and from micrometric measurements, it appears that several globular nebulae in each cluster are identical in size. In some cases it is possible to get complete super-position of their tracings. This fact suggests that globular nebulae may occur only in discrete sizes." This supposition was further borne out when angular diameters were corrected by cluster redshifts for distance effects. When the spherical galaxies in four clusters were compared, it was possible to identify three distinct sizes with the same linear diameters from cluster to cluster.

In addition to sizes of stars and galaxies, more recently Wilson, Cowan, and Burbidge have each published evidence for discrete distribu-

---

[4] Eggen, O. J.: *Astron. J. 60*, 407 (1955).

[5] Eggen, O. J., Sandage, A. R.: *Astrophys. J. 136*. 735 (1962).

[6] Page, T.: *Astrophys. J. 116*, 63 (1952).

[7] Wilson, A. G.: "Nebular Distances Determined From Their Diameters". Unpublished monograph read Mt. Wilson-C.I.T. Astrophysics Seminar, p. 25 (1950).

tions in the *redshifts* of clusters of galaxies, radio sources, and quasars [8]. It thus appears from the sizes of stars and galaxies and from the values of redshifts, that the possibility of discrete large scale distributions may be greater than has hitherto been suspected.

The most fundamental and obvious discrete distribution occuring in the cosmos is the *hierarchical structure* that differentiates aggregates of matter into stars, galaxies, clusters, and so on. These various levels of organization constitute a gross discretization effect, while the size and mass distributions among stars or among galaxies are but fine structure effects. The discretization phenomena to be discussed in this volume will be limited to these "fine structure" effects that are superimposed on the major structual discretizations. However, the fact that certain sub-levels in the inorganic world may attribute their discrete fine structure to relativistic effects, as predicted by the theory presented here, may also prove to be an important key to the origin of the gross-levels of cosmic structures.

The fine structure problem is a dual problem: The *existence* of discrete sub-levels must be established by suitable observational and statistical tests, and theoretical mechanisms consistent with existing physical theories that may account for the origin of such discrete distributions must be developed. Accordingly, this volume is divided into two parts. Part One develops a theory of physical isoplethic surfaces in states of geometric equilibrium within the discipline of General Relativity. An analysis of the environments of isoplethic surfaces is first given since the theory derives principally from existence conditions and these conditions hold only in certain environments. For galactic structures, the environmental problem is two-fold: the first is that of obtaining a cosmological environment that admits of anisotropies and inhomogeneities, in contrast to classical cosmology; the second is that associated with deriving averaging operators that preserve tensor structure so that the validity of the Einstein field equations for a stellar environment may be lifted to systems of field equations for the aggregates of stellar systems that comprise galaxies. An essential aspect of the theory is that no detailed assumptions are made concerning the momentum-energy complexes of galaxies, and accordingly, the theory is applicable to a wide class of physical equilibrium structures which satisfy certain specific conditions for the mathematical existence of solutions to the

---

[8] Wilson, A. G.: *Proc. Nat. Acad. Sci. 52*, 847 (1964); Wilson, A. G.: *Astron. J. 70*, 150 (1965); Wilson, A. G. In: *Proc. 14th International Astrophysical Symposium*, Vol. 41, pp. 125–129. Université de Liège 1967; Cowan, C. L.: *Astrophys. J. 154*, 15 (1968); Burbidge, G.: *Astrophys. J. 154*, 141 (1968); Burbidge, G. R., Burbidge, E. M.: *Nature 222*, 735 (1969).

Einstein field equations in the presence of jump discontinuities and the two conditions referred to as the isoplethic condition and the trace condition. The theory leads to predictions dealing with the morphology of isoplethic surfaces that constitute equilibrium constructs. These are illustrated in the text by figures of the predicted isoplethic surfaces for a wide range of the salient morphological parameters; one of which is discrete. This same discrete parameter that arises in the morphology is shown to characterize the diameters of elliptical galaxies, and in the case of *EO* galaxies leads to the conclusion that the diameters are proportional to the eigen-sequence $[n(n+1]^{1/2}$. Under the assumption that the factors of proportionality are the same from galaxy to galaxy in a given class, a specific discretization prediction results.

In Part Two, existing and new measurements of the diameters of early type elliptical galaxies are analyzed for evidence of the existence of discrete size distributions. Linear diameters of field galaxies are derived from observed angular diameters together with the assumption that distances are proportional to redshifts. Linear diameters of cluster galaxies are assumed to be proportional to their angular diameters. General discrete distributions and the specific $[n(n+1]^{1/2}$ distributions are investigated for both field and cluster galaxies using Monte Carlo methods to test the statistical significance of the results. Other possible discretized distributions such as those occurring among the observed redshifts of clusters of galaxies, radio sources, and other objects are investigated and the results compared with possible implications of the theory.

The studies leading to this volume have an origin that would have interested Wolfgang Pauli and Carl Jung – they would have detected the presence of what Jung defined as a "synchronicity"[9]. The authors of this tract, a mathematician and an astronomer, had independently developed theoretical and observational evidence for large scale discretization before they encountered each other on this subject in a hall of the RAND Corporation during November 1962. As already mentioned, Wilson had reported in 1950 the possibility that *EO* galaxies were quantized in size at a Mt. Wilson-Palomar Observatory Seminar. Prior to November of 1962, Edelen had developed for the case of quasi-spherical symmetry, a theory predicting that diameters of certain cosmic bodies, galaxies, possibly even stars and clusters, should be quantized or discretized. The authors had known each other for about a year before Edelen inquired one day whether there existed any astronomical data suggesting that galaxies occurred only in discrete sizes. Wilson was struck

[9] Jung, C. G.: "Synchronicity: An Acausal Connecting Principle", in *Collected Works*, Vol. 8, Ch. 7. New York: Pantheon Books 1960.

by this question because, had Edelen asked any other astronomer, the answer would undoubtedly have been negative. To his knowledge, Wilson was the only astronomer who had even suspected such an effect, but his own work of twelve years earlier had been put on the shelf for lack of sufficient observational evidence and in absence of theoretical reasons to support what appeared to be a set of discrete sizes among spherical galaxies. In this synchronistic encounter, it turned out that Edelen's prediction for diameters of spherical galaxies gave a very close fit to the preliminary Wilson data of twelve years before. Thus was born the study sponsored for two years by the RAND Corporation in theoretical and observational investigations into discretized cosmic phenomena. This book is the outgrowth of those studies.

The authors wish to acknowledge their indebtedness to the RAND Corporation for its support of the initial phases of this study and in particular to Drs. Stan Greenfield and Ted Harris; to the Mt. Wilson and Palomar Observatories for making available telescope time for a program to acquire direct images and redshifts of galaxies used to test discretization hypotheses. The authors wish to express their gratitude to Drs. Thornton Page, Gerard deVacouleurs, Jerry Neyman, Elizabeth Scott and Tracy Thomas for their interest in this problem and for many valuable discussions concerning observations, statistical tests and the general theory. Finally, the authors wish to express their appreciation for the interest and encouragement given to this study by the late Vaclav̌ Hlavatý and Abbé Lemaitre.

Contents

Prologue

The question of the existence of large scale discretized structures raises several basic theoretical and observational questions of an epistemological nature. It is the purpose of this prologue to formulate and discuss these questions and reference their later appearance in the text where they are developed in more detail. Therefore, in the original spirit of natural philosophy, we pause at the outset of this tract and make inquiry into the ideal state of affairs that can occur in physical discourse. The need for this is most apparent in subjects such as astronomy or atomic physics, for it is in these disciplines that the maximal contrast occurs between the natural scale of sensory awareness and the scales of the physical processes under examination.

In overview, this volume approaches the problem of large scale discretization through mathematical modeling of an entitation process of importance in the theory of levels: The derivation of the large scale properties of an aggregate when the local properties of the constitutent elements are known. The specific case chosen to explore the relations between large scale and local properties is a physical system whose elements possess momentum-energy tensors that obey the Einstein field equations of General Relativity. Under specified assumptions required to translate the perceptive operations of observers into the mathematical model, it is shown that certain parameters defining the allowable morphological forms and scales of the aggregate are *discretely distributed.* This case suggests applicability to the specific example of stars as elements and galaxies as aggregates, with the local properties being the gravitational field of stars and the large scale properties being the morphology of galaxies.

From this overview emerge several epistemological points that require amplification by way of support of the theoretical and observational formulations of Part One and Part Two. First, the distinction between continuous and discrete leads to the *problem of entitation*[1]. Second, the relation between an elemental entity and an aggregate entity involves the *problem of levels.* Finally, the identification of the

---

[1] Gerard, R. W.: Hierarchy, Entitation and Levels. In: *Hierarchical Structures.* Eds. Whyte, L. L., Wilson, A., and Wilson, D. New York: American Elsevier 1969.

objects and parameters of the theory with the observables and measurements of the laboratory raises some aspects of the *problem of modeling.*

The Problem of Entitation

A special case of the distinction between discrete and continuous will prove useful in the discussions of this and the following sections. We shall, therefore, as a preliminary, define a *binary* or discrete dichotomy and a *polar* or continuous dichotomy. Using *dichotomy* with the meaning of partitioning of a set into two subsets, we may have either an exhaustive and mutually exclusive A and $\bar{A}$ decomposition with every point of the set being mapped unequivically onto A or $\bar{A}$, or we may have a continuous distribution of the set between two poles such as north and south, plus and minus, with the location of the "equator" being arbitrary. The first decomposition we call binary; the second, we call polar. The binary dichotomy is usually associated with absolutes such as *existence* and *non-existence*, *truth* and *falsity*. The polar dichotomy is associated with sets for which we wish to preserve all or part of an ordering relation between elements, such as *rich* and *poor*, *young* and *old*.

The classical mode of thought and the basis of Aristotelian logic is through the reduction of our patterns of thinking onto a binary system. On the other hand, scientific experience indicates that it is better to consider the world as a polar system. Scientific thought recognizes the partial nature of our knowledge and that approximation underlies all our constructs. It follows that absolutes such as "true and false" are incompatible with the epistemological systems of creatures such as man that possess limitations of sensory experience, limitations in space and time, and limitations of knowledge. Since we can only know in part, we cannot meaningfully term our knowledge true or false. Rather, we can only rank the patterns we perceive by whether they are useful to us or whether they are in some way satisfying to us. These attributes are best measured along continuous scales that permit ordering, that is, they are *polar*. We may thus maintain that one representation of perception is more useful or that another is more elegant, but we may not maintain that a representation is true or false. Thus, in a choice between hypothesis (of representation) A and B we can only say hypothesis A is more elegant than B, or A fits the data better than B, or A is more economical than B, or A is simpler than B, but not that A is true and B is false. Unless the basis of knowledge is by some other process than the recognized processes of the scientific method, we are not justified in holding to the pre-scientific inheritances of the judgement of true or false. We thus re-name the so-called tests of validity and verification as tests of usefulness and satisfaction although they may still take the same statistical forms. Examination of tests of validity and verification shows that ultimately a subjective component, that demands either usefulness or some form of aesthetic satisfaction, is involved (e.g., a 3σ level of confidence). We propose not to go through the detour of self-delusion concerning truth and falsehood but rather to go to what is directly involved. Subjective concepts involving usefulness or aesthetic satisfaction are at the base of choices of hypotheses and hence the observer is very much part of all representations of perceptions.

It will be useful to reexamine another concept that is usually taken as binary – existence and nonexistence – and see whether it also may not be more appropriately considered as polar. The assumption of an absolute binary base for existence is a convention of thought that, while providing a useful simplification, denies the graded nature of the observer's perception and its relation to the concept of existence. The process of entitation naturally divides itself into the two complementary and interacting aspects of perception and conception. Although we shall discuss these two aspects sequentially, it must be born in mind that they are two complementary and inseparable aspects of human cognition.

The crux of the matter is that the assignment of existence is *a posteriori* with respect to perception, while nonexistence at any instant may be considered to consist of all elements of the universe of discourse that have not had an assignment of existence. Existence thus becomes a temporally ordered sequence of decisions and is thus polar with the "equator" being identified with the threshold of perception of the observer. The perception of a physical entity requires the *emergence* of at least one attribute of the entity from the substratum of sensory data where, by emergence of the entity, we mean intrusion through the observer's threshold of awareness. (Obviously, the threshold of different observers will differ depending both upon their previous experience and upon various augmentations of their sensory perceptions by means of instruments.) Now, an awareness threshold is binary, directly implying an on-off structure. The binary structure of the threshold of awareness must not be projected onto existence and nonexistence, however, nor clothed in unwarranted objectivity. The essential difference, that the threshold structure is observer-sensitive and binary, must be retained, while existence or nonexistence is independent of the observer and yet it becomes polar when related to the observer. We adopt this formulation of the ontological problem as the only meaningful one for scientific discourse and relegate other formulations of the existence, nonexistence problem to those concerned with other bases of knowledge.

The process of entitation not only involves *differentiation* of emergent entities from a sensory substratum through a perception sufficiently above the threshold level to trigger awareness, it also involves *discrimination* between all emerged entities through the cognitive processes of recognition of similarity and difference, of same and other. The process of discrimination consists of recognizing groupings within the set of perceptions, clustered according to the level of their interactions and correlations, and their identification with the sets that we call objects or entities.

A very important aspect of both the differentiation and discrimination processes is the bestowal of some form of *closure* on the entity. These closures may take the form of topological or spatial closure, temporal closure, interaction closure, etc.[2]. Certain aspects of physical entities (e.g., symmetries or quantitative measurements) are frequently abstracted and represented in mathematical conceptualizations. The closures associated with the psycho-physical entities may be represented by mathematical formulations of closure such as defined in group theory. Indeed, the success of mathematical models in representing the physical world is in no small part attributable to the fact that closure, basic to physical entitation, is also a basic property of mathematical conceptualizations. However, there is one important difference between the closures of psycho-physical entitation and those employed in mathematics. The closure of psycho-physical entities may be polar while the closures used in mathematical representations of the entities are binary. Thus in the physical domain, the differentiation "equator" between an entity and the substratum or background and the discrimination "equator" between two entities may be indefinite and somewhat arbitrarily located (as specifically arises in operational definitions of a galaxy, see Section 42), while in mathematical models, a closure interface providing a binary representation must be employed. This is an idealization peculiar to mathematical modeling like the geometric idealizations of planes without thickness and lines with no extension save in length. This idealization for the case of galaxies is further developed in Sections 16 and 21.

There is a remaining property of entities required by the dynamical view of perception. This is that the perception threshold level is determined not only by the intensity of stimuli, but also by the temporal persistence of the stimuli. Entities whose aspects persist in time are more readily perceived at a given threshold[3], and hence those aspects

---

[2] Wilson, A.: Closure, Entity and Level. In: *Hierarchical Structures*. Eds. Whyte, L. L., Wilson, A., and Wilson, D. New York: American Elsevier 1969.

[3] It must be pointed out, however, that persistence in time does not necessarily mean persistence at a constant amplitude. Awareness of a perception is most often enhanced through signal fluctuation.

of entities that have temporal persistence contribute to and aid perception. These aspects are those that are described in physical theory in terms of stability, equilibrium, or semiequilibrium (see Sections 13 and 14).

The Problem of Levels

A set of percepts that has a high degree of mutual interaction leads us to the concept of an *entity*. A grouping of sets possessing mutual interactions may lead to an aggregate entity and/or to the emergence of a *level*. Whether an aggregate entity or a new level emerges depends on the processes of aggregation. In general a non-linear combinatorial process that results in a whole greater than the sum of the parts produces what we usually term a higher level of organization. A formal discussion of the epistemological and ontological aspects of levels will not be undertaken here[4]; we shall instead focus on the question of how properties of an aggregate, as observed, are related to the observed properties of its component elements.

The idea that nature is structured in a series of levels of organization or integration, a *scala naturae*, is primarily neo-Platonic. This "ladder of nature" of Plotinus and other Hellenists may be termed a "hierarchical structure" in the sense of the higher levels governing the lower or the lower levels being derived from the top level which is the most perfect and complete. In this cosmology, the structures and organizations of the lower levels were either created "top down" or were fragments of the higher, more complete and complex organization. Mechanistic philosophy replaced this *scala naturae* with a single level universe populated with structures that evolve toward complexity from the simplest and most elemental "atoms"[5]. Mechanistic philosophy provides a "destructuring" principle in the second law of thermodynamics, but in the absence of a structuring or organizing principle[6] it cannot consistently account for the structures in the universe on the basis of the assumed initial condition of an unstructured "primeval atom" and

---

[4] Analyses of the problem of levels have been made by Woodger, J. H.: *Biology and Language*. Cambridge: Cambridge University Press 1952; Simon, H.: The Architecture of Complexity. *Proc. Amer. Philos. Soc. 106*, 467 (1962); Bunge, M.: The Metaphysics, Epistemology and Methodology of Levels. In: *Hierarchical Structures*. Eds. Whyte, L. L., Wilson, A., and Wilson, D. New York: American Elsevier 1969.

[5] Hawkins, D.: *The Language of Nature*. San Francisco: W. H. Freeman & Co. 1964; Bunge, M.: loc. cit.

[6] Morphogenesis of observed structure is usually "explained" on the basis of a series of highly improbable chance occurrences of various types. These may be random density fluctuations leading to the formation of galaxies or random groupings of molecules leading to living matter.

a "big bang". This unsatisfactory state of affairs has resulted recently in a reconsideration of a multi-level cosmos whose "top down" or "bottom up" information flows are unpostulated, remaining open to empirical investigation.

Several basic problems emerge in a multi-level universe. In addition to the question of the direction of flow of morphogenetic information, there is the question of the causes of plateaus or discrete levels in organization. Why, for example, do we observe only cosmic structures with masses in the exponential neighborhood of 10^{34} grams (stars) and 10^{45} grams (galaxies) and not structures occupying continuous range of mass. Finally, there is the question of central import to the present volume: How do we relate the properties of one level of organization to those of a higher or lower level of organization. It may well be that the solution of this question provides the key to the other questions.

We may formulate the inter-level property relation in terms of a set of basic descriptive properties $\{p_i\}$ that are sufficient to describe the individual or local properties of the i-th element of a system of A elements that constitute an aggregate. Let the combined effects of the p's be $\{P\} = \sum_A \{p_i\}$, "summed" over the elements of the aggregate with an appropriately defined operation of "addition", and let the "gestalt" interior properties of the differentiated aggregate be given by the set $\{\Pi\}$. We inquire in what instances will $\{\Pi\} = \{P\}$. In general, the set of gestalt properties may be derived from the sets of elemental properties by an appropriately defined superposition only in the case of continuous distributions of the elements comprising the aggregate. The gestalt properties of aggregates of discrete elements are not derivable, in general, from a linear aggregating rule or other superposition process, and there will generally exist an emergent quantity $E = \Pi - P \neq 0$. A specific example of this occurs in Part One in the derivation of the macroscopic properties from the "local" properties when the $\{p_i\}$ are the local momentum-energy tensors associated with each element of the aggregate (see Section 11).

However, the p's, P's and Π's are not usually known. Instead a set of "local observables", $\{o_i\}$, associated with each element and a set of "macroscopic observables", $\{\Omega\}$, associated with the aggregate, are measured. The problem becomes the inference of Π's and p's and their relation to each other from Ω's and o's on the basis of synthetic relations between the Ω's and o's that preserve internal consistency (see Sections 11 and 12). The relative spatial and temporal scales between the observer and the elements and aggregate observed, that are involved in the properties of Ω's and o's, suggests that the inferred physical laws are both scale and observer sensitive.

The Problem of Modeling

The subject of models, theories and constructs and their formulation and meaning has received extensive treatment in the literature[7]. This prologue does not pretend to summarize this subject, but only to single out and provide background for certain questions that arise in the formulation and interpretation of mathematical models of discrete aggregates. As pointed out in the first part of this prologue, distributions that may be *polar* in the physical world must be modeled as binary in order to be mathematically tractable. The problem in Part One is to construct a relativistic model of the physical system called a galaxy[8]. The problem in Part Two is to interpret the results of the mathematical operations carried out in Part One in terms of operationally defined parameters (see Sections 40 and 42). This becomes primarily an *identification* problem.

Observationally, a galaxy appears to be a concentration of stars, usually unresolved, having a luminosity distribution, $B(r)$, falling from a central maximum, B_0, according to a relation[9] closely approximated by the function,

$$B = B_0/(r + a)^2 ,$$

where a is a scale factor and r is the distance from the center. Galaxies have no perceptable outer boundary. Their apparent size depends on the exposure time. They extend with apparently monotonically decreasing luminosities until lost in the sky background or until they merge with another galaxy. These properties of galaxies make it both difficult to establish a boundary for theoretical purposes and to give size a meaningful operational definition. In Part One, the specific device used to establish a definite boundary of a galaxy as required in the theoretical treatment is to introduce a "jump discontinuity above a background field in some component of the momentum-energy tensor". This device, while precise and expedient mathematically, lacks the specificity to allow theoretical results governing the sizes of galaxies to be explicity identified

---

[7] See for example, Bunge, M.: *Scientific Research I: The Search for System*. Berlin-Heidelberg-New York: Springer 1967; Hawkins, D.: *The Language of Nature*. San Francisco: W. H. Freeman & Co. 1964; and Kuhn, T. S.: *The Structure of Scientific Revolutions*. Chicago: University of Chicago Press 1962.

[8] Observation and theory have independently identified *elliptical* galaxies as the objects of interest exhibiting discretization effects. Throughout this volume, except where explicitly otherwise remarked, "galaxy" is to mean "elliptical galaxy".

[9] Hubble, E.: *Astrophys. J. 71*, 231 (1930). Other empirical formulae are given by Vaucouleurs, G. de: General physical properties of external galaxies. In *Handbuch der Physik*, Vol. 53, 320, Berlin-Heidelberg-New York: Springer 1959; and Baum, W. A.: *Pub. Astron. Soc. Pacific 67*, 328 (1955).

with any operationally defined diameter. Accordingly we encounter an *a posteriori* identification problem. If a set of observables is candidate for the role of physical counterparts to a set of theoretical objects, then the two sets must be dimensionally and quantitatively the same; or in the absence of the specification of an absolute scale by the theory, the two sets, normalized to their largest or their smallest member, must be dimensionless and quantitatively the same. If diameters of several galaxies, however operationally defined, are proportional, and therefore all exhibit the same relative properties, then any convenient set of diameters may be used to test the theory. If, on the other hand, diameters measured by different operations are not proportional, then the theory becomes untestable unless the particular diameter corresponding to the theory can be isolated. Since the theory is not sufficiently specific for this purpose, its applicability to galaxies must depend on another property of galaxies that is not considered by the theory, namely, proportionality of diameters corresponding to different isophotes. If the above luminosity distribution law of Hubble's may be assumed, then for $r/a \gg 1$ (i.e., for diameters defined by isophotes of a small percentage of the central intensity, as is required by the theory) proportionality exists and the precise specification of diameters is not important. Further aspects of the identification problem are discussed in Section 42.

We have used this Prologue to introduce, within a general context, certain epistemological problems before their particularizations are treated in Part One or Part Two. In this way we have attempted to assure adequate exposure of the difficulties these problems pose, even if they have not in every case been fully treated or resolved. It is hoped that through this exposure more attention will be given to these and similar epistemological problems, which grow in both importance and consequence as a direct result of the ever-increasing information within the channels of science.

PART ONE

Theoretical Considerations

Unless noted to the contrary, we follow the notation, the definitions of symbols, and the definitions of operations given by Schouten[1]. Wherever possible, we have used Schouten as a standard reference. Since several different spaces will be used simultaneously, we have found it useful to distinguish tensor-valued functions on the various spaces by use of different types of index alphabets.

Capital Roman indices, A, B, C, ... have the range 0, 1, 2, 3 and signify tensor-valued functions on a 4-dimensional, hyperbolic-normal metric space.

Lower case Roman indices, i, j, k, ... have the range 1, 2, 3 and signify tensor-valued functions on a 3-dimensional Riemannian manifold. These Riemannian manifolds are usually 3-dimensional spacelike slices of a 4-dimensional, hyperbolic-normal metric space.

Capital Greek indices, Γ, Σ, Λ, ... have the range 0, 1, 2 and signify tensor-valued functions on a 3-dimensional, hyperbolic-normal metric space. Such spaces are usually 3-dimensional timelike hypersurfaces of a 4-dimensional, hyperbolic-normal metric space.

Lower case Greek indices, α, β, γ, ... have the range 1, 2 and signify tensor-valued functions on a 2-dimensional Riemannian manifold. These 2-dimensional manifolds are usually space-like slices of the 3-dimensional hyperbolic-normal metric spaces that are signified by capital Greek indices.

Lower case German gothic indices $\mathfrak{a}$, $\mathfrak{b}$, $\mathfrak{c}$, ... have the range 0, 1, 2, 3. These indices are used solely as labeling indices and signify a specific vector or tensor from a collection of such quantities that forms a basis for the tensor algebra or of a subalgebra of the tensor algebra. *German gothic indices are not tensor indices.*

The summation convention is assumed throughout this Part. It is to be applied to all five types of indices enumerated above. The range of each implied summation is determined by the range of the type of index involved.

---

[1] Schouten, J. A.: *Ricci-Calculus*, 2nd ed. Berlin-Göttingen-Heidelberg: Springer 1954.

CHAPTER I

The Environment for Cosmic Structures

The purpose of this Chapter is to obtain a description of the galactic environment. The question of the environment is of more importance in the theory given in these pages than is usually the case, for many of the arguments hinge on existence conditions and these conditions obtain only in certain environments. The key element is the form and structure of the momentum-energy tensor, and in particular, the deviations of the momentum-energy tensor of the galactic environment from that of a perfect fluid.

There is a twofold environmental problem to be examined. The first aspect obtains from the consideration of the inhomogeneities and anisotropies of the enveloping cosmological environment and is studied in Part A. The environmental description thus obtained allows us to examine galaxies in terms of their emergence from the cosmological structure as a background field. The second aspect stems from the more subtle difficulties involved in attempts to infer a macroscopic structure from a local structure; in particular, that of obtaining the defining equations of an aggregate when the constituents of the aggregate satisfy the Einstein field equations. This part of the problem is taken up in Part B. An aggregation or observation operator is shown to be defined on the tensor algebra. This operator and its associated commutator with the operators involved in the Einstein field equations gives a partial resolution of the aggregation problem and a description of the environment from the standpoint of internal consistency. The two aspects of the problem are mutually complementary – yielding descriptions of the environment from the general (cosmological) and from the particular (local) – and coalesce in giving descriptions of the momentum-energy tensors that exhibit similar deviations from that of a perfect fluid.

A. The External or Cosmological Enveloping Environment

An assumed cosmological structure acts as an environment within which the study of galaxies and similar structures is to be pursued. With this view in mind, it is evident that we must develop cosmological models that allow for agglomerations of matter – for certainly we can not use the classical models that preclude the very existence of the inhomogeneities and anisotropies that are requisite for galactic structures. The inhomogeneous and anisotropic cosmological models that are presented in the following pages will be used as the external enveloping environment for the study of galactic structures that is developed in succeeding Chapters.

There is also an overriding problem that must be solved if the currently accepted correlations between the predictions of the classical cosmological models and the observables in cosmology are to be preserved. Most observables that are available for cosmological investigations stem from the agglomerations of matter in the form of galaxies, clusters of galaxies, and possibly super clusters of galaxies (second order clusters). The classical models, however, assume a homogeneous and isotropic spatial distribution of matter and pressure within the context of the Einstein field equations. It is thus necessary to construct inhomogeneous models that are consistent with the observed agglomerations of matter, and then to show that the basic effects predicted by the classical models are changed in only a negligible manner or in a precisely predicted and computed fashion by the actual agglomerations of matter that must be used to test the theory. A number of analyses and methods have been aimed in this direction in recent years [1], but significant work still remains to be done. No attempt is made in these pages to give a resolution of the full description of the observables and their dependence on the agglomerations of matter, since this would take us well outside of the intended scope of this work.

1. The Classical Models

We collect here those results from the classical cosmological models that are required for purposes of reference and comparison. The reader is referred to Robertson [2] and to Heckmann and Schücking [3] for the details.

[1] Bonnor, W. B.: *Monthly Notices Roy. Astron. Soc. 117,* 104 (1957), Newtonian perturbations; Irvine, W.: *Ann. Physics 32,* 322 (1965), Newtonian approximation in relativistic models; Lifshitz, E. M.: *J. Phys. U.S.S.R. 10,* 116 (1946); Lifshitz, E. M., Khalatnikov, I. N.: *Advances in Phys. 12,* 185 (1963), Metric perturbations of relativistic models; Hawking, S. W.: *Astrophys. J. 145,* 544 (1966), Perturbations of the conformal tensor; Chandrasekhar, S.: *Astrophys. J. 142,* 1488 (1965), Post Newtonian approximations; Edelen, D. G. B.: *Nuovo Cimento 43 A,* 1095 (1966), Metric and energy perturbations of conformally static spaces; Edelen, D. G. B.: *Nuovo Cimento 55 B,* 155 (1968), Conformally homogeneous models.

[2] Robertson, H. P.: *Rev. Modern Phys. 5,* 62 (1933).

[3] Heckmann, O., Schücking, E.: Relativistic Cosmology. In: *Gravitation: An Introduction to Current Research,* Chapter XI. New York: John Wiley & Sons 1962.

Let $\mathscr{E}$ denote the class of Einstein-Riemann spaces[4] admissible in classical cosmological models and let $g_{\mathrm{AB}}(x^{\mathrm{K}})$ denote the components of the metric tensor of an element of $\mathscr{E}$. We then have

$$\mathrm{d}s(g)^2 = g_{\mathrm{AB}}\,\mathrm{d}x^{\mathrm{A}}\,\mathrm{d}x^{\mathrm{B}} \overset{*}{=} \mathrm{d}t^2 - R(t)^2\,\mathrm{d}u(L)^2\,, \tag{1.1}$$

where $\overset{*}{=}$ denotes an evaluation in a coordinate system (a$*$-coordinate system) in which the matter in the model universe is at rest (comoving coordinates), the value of the speed of light *in vacuo* is unity, and

$$\begin{aligned} \mathrm{d}u(L)^2 &= L_{ij}(x^k)\,\mathrm{d}x^i\,\mathrm{d}x^j \\ &\overset{*}{=} (1 + kr^2/4)^{-2}((\mathrm{d}x^1)^2 + (\mathrm{d}x^2)^2 + (\mathrm{d}x^3)^2)\,, \\ (r^2 &\overset{*}{=} (x^1)^2 + (x^2)^2 + (x^3)^2)\,, \end{aligned} \tag{1.2}$$

is the line element on the three-dimensional space $\mathscr{S}$ of constant curvature, $k = 1, 0, -1$[5]. The "perfect fluid" momentum-energy tensor associated with any element of $\mathscr{E}$ is given by

$$T_{AB} = (\rho + p)\,W_A W_B - p g_{AB}\,, \tag{1.3}$$

where

$$W^A \overset{*}{=} \delta^A_0 \qquad (W^A W^B g_{AB} = 1) \tag{1.4}$$

is the velocity vector and ρ and p are the density and (Poincaré) pressure, respectively. If $G_{AB}(g)$ denote the components of the Einstein tensor formed from $\{g_{AB}\}$, the physics and the geometry are related by the Einstein field equations

$$G_{AB}(g) = \kappa T_{AB} + \Lambda g_{AB}\,, \tag{1.5}$$

where Λ is the cosmological constant[6]. When a $*$-coordinate system is used, these equations yield

$$\begin{aligned} \kappa\rho &\overset{*}{=} -\Lambda + 3(k + \dot{R}^2)R^{-2}\,, \qquad (\dot{R} = \mathrm{d}R/\mathrm{d}t)\,, \\ \kappa p &\overset{*}{=} \Lambda - 2\ddot{R}R^{-1} - (k + \dot{R}^2)R^{-2}\,, \\ 0 &\overset{*}{=} R\dot{\rho} + 3\dot{R}(\rho + p)\,, \end{aligned} \tag{1.6}$$

and the nonvanishing Christoffle symbols of the second kind are

$$\Gamma^i_{jk}(g) \overset{*}{=} \Gamma^i_{jk}(L)\,, \qquad \Gamma^0_{ij}(g) \overset{*}{=} R\dot{R}\mathrm{L}_{ij}\,, \qquad \Gamma^i_{0j}(g) \overset{*}{=} \dot{R}R^{-1}\delta^i_j\,. \tag{1.7}$$

---

[4] Einstein-Riemann spaces are hyperbolic-normal metric spaces in which Einstein's field equations of general relativity hold.

[5] A kinematic development that leads to (1.1) and (1.2) without the assumption of comoving coordinates is given by Edelen, D. G. B.: *Arch. Rational Mech. Anal.* *25*, 159 (1967).

[6] See Appendix A for the definition and the sign convention for the curvature tensor and the Einstein tensor used here.

All elements of $\mathscr{E}$ are spherically symmetric and conformally flat, and hence they admit the maximal number of conformal motions, namely fifteen[7]. By a conformal motion, we mean a (generating) vector field $\{\xi^A(x^K)\}$ such that

$$£(\xi)(g_{AB}) = \nabla_A \xi_B + \nabla_B \xi_A = \varphi(x^K) g_{AB}$$

for some bounded scalar function $\varphi(x^K)$, where $£(\xi)$ denotes the operation of Lie differentiation[8] with respect to the vector field $\{\xi^A\}$. The work of Takeno cited above gives a full and detailed account of the conformal motions as well as the groups of motions (Killing fields) admitted by the various elements of $\mathscr{E}$.

With Eqs. (1.4) and (1.7) and the definition of the expansion parameter, $\theta \equiv \nabla_A W^A$, and the rotation

$$\omega_{AB} = \tfrac{1}{2}(\delta^C_A - W_A W^C)(\delta^D_B - W_B W^D)(\nabla_C W_D - \nabla_D W_C),$$

we have

$$\theta \overset{*}{=} 3\dot{R}R^{-1}, \qquad \omega_{AB} \overset{*}{=} 0. \tag{1.8}$$

Hence, the Hubble parameter, $H = H(t)$, is given by

$$H \overset{*}{=} \dot{R}R^{-1} \overset{*}{=} \theta/3. \tag{1.9}$$

2. Inhomogeneity as a Similarity Construct

In view of the fundamental equivalence of geometry and physics stated by the Einstein field equations, physical inhomogeneities, such as the agglomeration of matter into galaxies and clusters of galaxies, imply spatial geometric inhomogeneities. The simplest manner of introducing such inhomogeneities, and one that is consistent with the idea that the inhomogeneities can be viewed as "bumps", so to speak, on a homogeneous substratum, is that in which the spatial geometry of the inhomogeneities is similar to the geometry of the homogeneous substratum. The factor of proportionality obtained in stating the similarity as an equality would then describe the inhomogeneities through the variation in the values of the proportionality factor from point to point. If there are spatial geometric inhomogeneities, however, the corresponding physical inhomogeneities imply an inhomogeneity in the energy density, and this in turn would imply inhomogeneities in the proper time rates. Since we must thus consider both spatial and temporal inhomogeneities, we are led to construct inhomogeneous cosmological

[7] Takeno, H.: *The Theory of Spherically Symmetric Space-Times,* Scientific Reports of the Research Institute for Theoretical Physics, Hiroshima University, No. 3 (1963).

[8] Schouten: p. 102ff.

models that are *conformally homogeneous:* the 4-dimensional metric geometry of the inhomogeneous model is similar to the 4-dimensional metric geometry of the homogeneous substratum. The factor of proportionality (the conformal factor) then describes the effects of the inhomogeneities.

Above and beyond the arguments of both mathematical and physical simplicity, conformal homogeneous models possess a unique property that singles them out from all other possible inhomogeneous models. It is known that the complete curvature tensor of a metric space can be decomposed uniquely into the sum of two tensors, one of which is the Weyl conformal curvature tensor and the other is a tensor that is uniquely determined by the metric tensor and the Ricci tensor (by the metric tensor and the momentum-energy tensor)[9]. Now, the Weyl tensor is that part of the curvature tensor that is not determined locally by the matter (by the momentum-energy tensor) and may thus be viewed as representing the "free gravitational field"[10]; further, the Weyl tensor is the unique curvature invariant under conformal changes in the metric tensor[11]. It thus follows that conformally homogeneous models describe inhomogeneous destributions of matter whose free gravitational fields are identical with the free gravitational fields of the corresponding homogeneous models; namely, identically vanishing since the homogeneous models are conformally flat[12].

Since the homogeneous models are conformally flat, they admit the full fifteen-parameter group of conformal motions, and this fact, to a great extent, determines a majority of the predictions of observables in the classical models[13]. In this respect, the conformally homogeneous models admit the same fifteen-parameter conformal group of motions as admitted by the classical models and they are the *unique* models whose generating vectors of the fifteen-parameter conformal group are the same as those for the classical models[14]. The essential difference between the classical models and the conformally homogeneous models is that the fifteen-parameter conformal group of the classical models admits

---

[9] Schouten: p. 306, Eqs. (5.17), (5.18).

[10] Jordan, P., Ehlers, J., Knudt, W.: *Abh. Akad. Wiss. Mainz.*, No. 2 (1960).

[11] Schouten: p. 306.

[12] Note should be taken that there is a basic philosophical problem involved here: how does one build a model that represents the whole universe and claim any validity for it if the model admits a free gravitational field that is independent of the matter in the universe? Unless extreme care is used, free gravitational field universes amount to a direct contradiction of terms and can be salvaged only by means barely shy of pontification.

[13] Hawking, S. W.: *Astrophys. J. 145*, 544 (1966); Sandage, A. R.: *Astrophys. J. 133*, 355 (1961).

[14] Takeno, H.: *op. cit.*

a six-parameter subgroup that forms a group of metric motions (Killing fields) that is isomorphic to the six-parameter group of metric motions in Euclidean 3-dimensional space. This six-parameter group of metric motions is what gives rise to the spatial homogeneity and isotropy of the classical models. The conformally homogeneous models, on the other hand, will not admit a six-parameter group of metric motions in general. From the group-theoretic point of view, it thus follows that the conformally homogeneous models preserve the conformal group structure while destroying just the metric motions that give rise to isotropy and homogeneity. Conformally homogeneous models may thus be expected to preserve most of the agreement between the predictions of the classical theory and the cosmological observables that do not depend on the inhomogeneities, and yet afford the freedom to model the actual agglomerations of matter that are found in the universe.

The construction of inhomogeneous cosmological models is not just a formal nicety; rather, it is an epistomological necessity, for it is only when it is shown that the predictions of homogeneous and inhomogeneous models are approximately the same that any confidence can be placed in the correlations between the predictions of the homogeneous models and the cosmological observables. This follows from the simple fact that the observable quantities pertain to just those agglomerations of matter in the forms of galaxies and clusters of galaxies that are precluded by the assumptions of isotropy and homogeneity that underly the classical models.

Let $\mathscr{C}$ denote the class of Einstein-Riemann spaces that are conformally equivalent to elements of $\mathscr{E}$, and let $h_{AB}(x^K)$ denote the components of the metric tensor of such spaces. We then have

$$h_{AB}(x^K) = \exp\{\psi(x^K)\}\, g_{AB}(x^K), \tag{2.1}$$

where

$$\mathrm{d}s(h)^2 = h_{AB}\,\mathrm{d}x^A\,\mathrm{d}x^B = \exp\{\psi\}\,\mathrm{d}s(g)^2 \tag{2.2}$$

is the fundamental metric form on elements of $\mathscr{C}$ and ψ is the *conformal coefficient*. The structure of such spaces is well known and is given in Appendix C. Since there are two different metric tensors, $\boldsymbol{g}$ and $\boldsymbol{h}$, at our disposal, care must be exercised in order to distinguish which tensor is used to raise and lower tensor indices. Briefly, the conventions to be used are as follows and are spelled out in detail in Appendix C: if a tensor index is raised or lowered by means of the metric tensor $\{h_{AB}\}$, a dot is inserted in the original place of the index; raising and lowering of tensor indices by means of the metric tensor $\{g_{AB}\}$ follow the usual conventions: $\underline{\nabla}$ is used to denote covariant differentiation based on the metric tensor

$\{h_{AB}\}$; ∇ is used to denote covariant differentiation based on the metric tensor $\{g_{AB}\}$. With this notation, we have[15]

$$\begin{aligned} \Gamma^A_{BC}(h) &= \Gamma^A_{BC}(g) + \tfrac{1}{2}(\delta^A_B \partial_C \psi + \delta^A_C \partial_B \psi - g^{AD} g_{BC} \partial_D \psi), \\ G_{AB}(h) &= G_{AB}(g) - \nabla_A \nabla_B \psi + \tfrac{1}{2} \partial_A \psi \partial_B \psi \\ &\quad + g_{AB} g^{CD} (\nabla_C \nabla_D \psi + \tfrac{1}{4} \partial_C \psi \partial_D \psi), \end{aligned} \tag{2.3}$$

where $\{G_{AB}(h)\}$ denotes the Einstein tensor formed from $\{h_{AB}\}$. The field equations for elements of $\mathscr{C}$ are given by

$$G_{AB}(h) = \kappa\, 'T_{AB}^{\cdot\cdot} + {}'\Lambda h_{AB}, \tag{2.4}$$

where $\{'T_{AB}^{\cdot\cdot}\}$ and $'\Lambda$ are the momentum-energy tensor and the cosmological constant, respectively, for the conformally homogeneous models. It now follows from (1.3), (1.5), (2.1), (2.3) and (2.5) that the field Eqs. (2.4) are equivalent to the system of equations

$$\begin{aligned} \kappa\, 'T_{AB}^{\cdot\cdot} &= \kappa(\rho + p) W_A W_B - \nabla_A \nabla_B \psi + \tfrac{1}{2} \partial_A \psi \partial_B \psi \\ &\quad + \{\Lambda - {}'\Lambda e^{\psi} - \kappa p + g^{CD}(\nabla_C \nabla_D \psi + \tfrac{1}{4} \partial_C \psi \partial_D \psi)\} g_{AB}. \end{aligned} \tag{2.5}$$

3. The Momentum-Energy Tensor

It would be most convenient if we could assume that the momentum-energy tensor $\{'T_{AB}^{\cdot\cdot}\}$ associated with conformally homogeneous models is representative of a perfect fluid, for this assumption appears essential for the physical interpretations in the corresponding classical models. The perfect fluid momentum-energy tensor, (1.3), is so structured, however, that it has a simple eigenvalue yielding a time-like eigenvector (the velocity vector $\boldsymbol{W}$) and an eigenvalue of multiplicity three, yielding an isotropic 3-dimensional vector space of space-like eigenvectors. A glance at (2.5) shows that the terms $\nabla_A \nabla_B \psi$ and $\nabla_A \psi \nabla_B \psi$ will, in general, preclude such an eigenvalue and eigenvector structure for conformally homogeneous models. We must accordingly take the more general form

$$'T^{AB} = ('\rho + {}'p) U^A U^B - {}'p h^{AB} + Q^{AB}, \tag{3.1}$$

where $\{Q^{AB}\}$ constitute the components of a symmetric tensor field that is to be determined. In view of the term $'p h^{AB}$ in (3.1), there is no loss of generality in assuming that $\boldsymbol{Q}$ is trace free:

$$Q^{AB} h_{AB} = Q^A{}_{A}^{\cdot} = Q^A_A e^{\psi} = 0. \tag{3.2}$$

---

[15] See Appendix C for the details.

If $'\rho$ is to be the eigenvalue corresponding to the eigenvector $\boldsymbol{U}$, we require

$$Q^{AB} h_{BC} U^C = Q^{AB} U_{\dot{B}} = Q^{AB} U_B e^{\psi} = 0. \tag{3.3}$$

With the eigenvector character of $\boldsymbol{U}$ thus guaranteed, there is no loss of generality in normalizing $\boldsymbol{U}$ so that it is a unit vector. This unit vector must be time-like if $'\rho$ is to be associated with the mass, and hence we have

$$U^A U^B h_{AB} = U^A U_{\dot{A}} = U^A U_A e^{\psi} = 1. \tag{3.4}$$

Under these conditions, $'\rho$ and $\boldsymbol{U}$ can be identified with the total rest energy and the velocity vector field of energy transport in elements of $\mathscr{C}$, respectively[16]. Conformally homogeneous models thus allow consideration of more general dynamical processes than those exhibited by a perfect fluid, in fact, they demand it. This is to the advantage of the theory, however, in view of the scope and energy of galactic magnetic fields[17] that are neglected in the classical models. The occurrence of the tensor $\boldsymbol{Q}$ in (3.1), in addition to the terms that constitute the perfect fluid momentum-energy tensor, will have significent bearing later on, and will serve to describe the anisotropy since $Q_{AB} \neq 0$ precludes an isotropic 3-space of eigenvectors of $\{'T^{AB}\}$.

Since both $\boldsymbol{W}$ and $\boldsymbol{U}$ are time-like vector fields with respect to their corresponding metric tensors and conformal transformations such as (2.1) map null cones onto null cones, we may write

$$U^A = \lambda W^A + V^A, \qquad V^A W^B h_{AB} = V^A W_A e^{\psi} = 0. \tag{3.5}$$

where λ and $\boldsymbol{V}$ are to be determined. With this decomposition, we relate the velocity vectors of the homogeneous and the conformally homogeneous models and thereby interpret $\boldsymbol{V}$ as the *dispersion vector* of the conformally homogeneous model with respect to the corresponding velocity vector field of the homogeneous model. In this way we can make physical sense of the conformally homogeneous models even though there are, in general, no isotropic comoving coordinate systems on elements of $\mathscr{C}$. Use of the normality condition $U^A U_{\dot{A}} = 1$ with (1.4), (2.1) and (3.5) gives us

$$\lambda = \sqrt{e^{-\psi} + V^2}, \qquad V^2 = -g_{AB} V^A V^B \geq 0, \tag{3.6}$$

where the sign of the radical is determined by the requirement that both $\boldsymbol{W}$ and $\boldsymbol{U}$ point into the future null cone.

---

[16] Edelen, D. G. B.: *Nuovo Cimento 30*, 292 (1963).

[17] Woltjer, L.: Galactic Magnetic Fields. In: *The Structure and Evolution of Galaxies*. Ernest Solvay ed. London: Interscience Publishers 1965.

A combination of the above results now gives us the following form for $'T^{\cdot\cdot}_{AB}$:

$$\begin{aligned}'T^{\cdot\cdot}_{AB} &= e^{2\psi} Q_{AB} - 'p e^{\psi} g_{AB} \\ &\quad + ('\rho + 'p) e^{2\psi} \{\lambda^2 W_A W_B + V_A V_B + \lambda(V_A W_B + V_B W_A)\}\,.\end{aligned} \tag{3.7}$$

4. The Governing Equations

We can now obtain the governing equations by eliminating $\{'T^{\cdot\cdot}_{AB}\}$ between (2.5) and (3.1), or between (2.5) and (3.7). If this is done, and we use the abreviations

$$\Box^2 \psi = g^{AB} \nabla_A \nabla_B \psi\,, \qquad \chi = g^{AB} \partial_A \psi \partial_B \psi\,, \tag{4.1}$$

the results are

$$\begin{aligned}&\kappa('\rho + 'p) e^{2\psi} U_A U_B - \kappa 'p e^{\psi} g_{AB} + \kappa e^{2\psi} Q_{AB} \\ &\quad = \kappa(\rho + p) W_A W_B - \nabla_A \nabla_B \psi + \tfrac{1}{2} \partial_A \psi \partial_B \psi \\ &\qquad + \{\Box^2 \psi + \tfrac{1}{4}\chi + \Lambda - \kappa p - '\Lambda e^{\psi}\} g_{AB}\,,\end{aligned} \tag{4.2}$$

or

$$\begin{aligned}&\kappa('\rho + 'p) e^{2\psi} \{\lambda^2 W_A W_B + V_A V_B + \lambda(V_A W_B + V_B W_A)\} \\ &\quad - \kappa 'p e^{\psi} g_{AB} + \kappa e^{2\psi} Q_{AB} \\ &\quad = \kappa(\rho + p) W_A W_B - \nabla_A \nabla_B \psi + \tfrac{1}{2} \partial_A \psi \partial_B \psi \\ &\qquad + \{\Box^2 \psi + \tfrac{1}{4}\chi + \Lambda - \kappa p - '\Lambda e^{\psi}\} g_{AB}\,.\end{aligned} \tag{4.3}$$

When Eqs. (1.6) are combined with either of the above sets, the method of derivation shows that the combined systems (1.6) and (4.2) or (1.6) and (4.3) are equivalent to the Einstein field equations in $\mathscr{C}$. Note that the metric tensor $\boldsymbol{g}$ is assumed to be known from the classical model (see (1.1) and (1.2)) and hence its ten independent components are not to be counted as unknown variables in the above system. The same observation holds for the quantities ρ, p and Λ.

We now use the $*$-coordinate systems introduced in Section 1 to rewrite the system (4.3) in an alternative and somewhat more useful form. A straightforward calculation based on (1.4), (3.5) and (3.6) gives

$$V^0 \stackrel{*}{=} V_0 \stackrel{*}{=} 0\,, \qquad V_i \stackrel{*}{=} -R^2 L_{ij} V^j\,, \qquad V^2 \stackrel{*}{=} R^2 L_{ij} V^i V^j\,, \tag{4.4}$$

and hence we have

$$U_0 \stackrel{*}{=} \lambda e^{\psi}\,, \qquad U_i \stackrel{*}{=} -R^2 e^{\psi} L_{ij} V^j\,, \qquad U^0 \stackrel{*}{=} \lambda\,, \qquad U^i \stackrel{*}{=} V^i\,. \tag{4.5}$$

The condition that $\boldsymbol{Q}$ admit $\boldsymbol{U}$ as a null vector, together with (3.5) and the above results show that

$$\begin{aligned} Q^{00} &\stackrel{*}{=} \lambda^{-2} Q^{ij} V_i V_j, & Q^{0i} &\stackrel{*}{=} -\lambda^{-1} Q^{ij} V_j, \\ Q_{00} &\stackrel{*}{=} Q^{00}, & Q_{0i} &\stackrel{*}{=} \lambda^{-1} R^2 L_{ij} Q^{jk} V_k, \end{aligned} \tag{4.6}$$

while the trace-free condition on $\boldsymbol{Q}$ gives

$$Q^{00} \stackrel{*}{=} R^2 L_{ij} Q^{ij} \tag{4.7}$$

and hence we have (from (4.6))

$$(R^2 L_{ij} - \lambda^{-2} V_i V_j) Q^{ij} \stackrel{*}{=} 0 . \tag{4.8}$$

When (4.1) is combined with the definition of covariant differentiation, (1.7), and (2.3), it is easily seen that

$$\begin{aligned} \Box^2 \psi &\stackrel{*}{=} \partial_0 \partial_0 \psi - R^{-2} L^{ij} (\partial_i \partial_j \psi - \Gamma^k_{ij}(L) \partial_k \psi - R \dot{R} L_{ij} \partial_0 \psi), \\ \chi &\stackrel{*}{=} (\partial_0 \psi)^2 - R^{-2} L^{ij} \partial_i \psi \partial_j \psi . \end{aligned} \tag{4.9}$$

The system (4.3) then becomes

$$\begin{aligned} \kappa\{('\rho + 'p) e^{2\psi} \lambda^2 - \rho - e^{\psi} 'p\} + '\Lambda e^{\psi} - \Lambda + \kappa \lambda^{-2} Q^{ij} V_i V_j e^{2\psi} \\ \stackrel{*}{=} -\partial_0 \partial_0 \psi + \tfrac{1}{2} (\partial_0 \psi)^2 + \Box^2 \psi + \tfrac{1}{4} \chi \end{aligned} \tag{4.10}$$

for $A = B = 0$,

$$\begin{aligned} -\kappa('\rho + 'p) e^{2\psi} \lambda R^2 V^j L_{ij} - \kappa \lambda^{-1} R^4 e^{2\psi} L_{ij} Q^{jk} L_{km} V^m \\ \stackrel{*}{=} -\partial_0 \partial_i \psi + (\dot{R} R^{-1} + \tfrac{1}{2} \partial_0 \psi) \partial_i \psi \end{aligned} \tag{4.11}$$

for $A = 0$ and $B = i$, and

$$\begin{aligned} \kappa('\rho + 'p) e^{2\psi} R^4 L_{ki} L_{jm} V^k V^m + \kappa('p e^{2\psi} - p) R^2 L_{ij} \\ + (\Lambda - '\Lambda e^{\psi}) R^2 L_{ij} + \kappa R^4 e^{2\psi} L_{ik} L_{jm} Q^{km} \\ \stackrel{*}{=} -\partial_i \partial_j \psi + \Gamma^k_{ij}(L) \partial_k \psi + R \dot{R} L_{ij} \partial_0 \psi \\ + \tfrac{1}{2} \partial_i \psi \partial_j \psi - R^2 L_{ij} (\Box^2 \psi + \tfrac{1}{4} \chi) \end{aligned} \tag{4.12}$$

for $A = i$ and $B = j$. When the value of λ given by (3.6) is substituted into (4.10), we obtain the equivalent and more useful equality

$$\begin{aligned} \kappa\{'\rho e^{\psi} - \rho + V^2('\rho + 'p) e^{2\psi}\} + '\Lambda e^{\psi} - \Lambda + \kappa \lambda^{-2} e^{2\psi} Q^{ij} V_i V_j \\ \stackrel{*}{=} -\partial_0 \partial_0 \psi + \tfrac{1}{2} (\partial_0 \psi)^2 + \Box^2 \psi + \tfrac{1}{4} \chi . \end{aligned} \tag{4.13}$$

This system of equations, together with the system (1.5), although equivalent to the field equations, gives direct relations between the salient physical quantities of the conformally homogeneous models, the conformal coefficient and its derivatives, and the known quantities of the corresponding homogeneous models.

Before ending this Section, we note several additional results that will be useful later. If we multiply (4.3) by $\{g^{AB}\}$ and sum on the repeated indices, a little work and our previous results give us

$$\kappa('\rho - 3'p) e^{\psi} = \kappa(\rho - 3p) + 3(\Box^2 \psi + \tfrac{1}{2} \chi) + 4(\Lambda - '\Lambda e^{\psi}) . \tag{4.14}$$

A multiplication of (4.3) by $\{U^A\}$ and then by $\{U^B\}$, together with several algebraic manipulations, lead to the following results:

$$\kappa\,'\rho = -'\Lambda + (\kappa\rho + \Lambda)e^{-\psi} + \kappa V^2(\rho + p) + e^{-\psi}(\Box^2\psi + \tfrac{1}{4}\chi) - U^A U^B(\nabla_A\nabla_B\psi - \tfrac{1}{2}\partial_A\psi\,\partial_B\psi), \tag{4.15}$$

$$3\kappa\,'p = 3\,'\Lambda - 3\Lambda e^{-\psi} + 3\kappa p e^{-\psi} + \kappa V^2(\rho + p) - e^{-\psi}(2\Box^2\psi + \tfrac{5}{4}\chi) - U^A U^B(\nabla_A\nabla_B\psi - \tfrac{1}{2}\partial_A\psi\,\partial_B\psi), \tag{4.16}$$

$$(\nabla_A\nabla_B\psi - \tfrac{1}{2}\partial_A\psi\,\partial_B\psi)U^A = \kappa(\rho + p)\lambda W_B + \{\Lambda - '\Lambda e^{\psi} - \kappa p + \Box^2\psi + \tfrac{1}{4}\chi - \kappa\,'\rho e^{\psi}\}U_B. \tag{4.17}$$

Eqs. (4.11), (4.12) and (4.17) may be viewed as defining the state of anisotropy of conformally homogeneous models since they serve to determine the quantities $\{Q^{AB}\}$.

5. The Dispersion Vector

The decomposition given by (3.5) allows us to represent the velocity vector $\boldsymbol{U}$ of the conformally homogeneous model in terms of the velocity vector $\boldsymbol{W}$ of the corresponding classical models and the dispersion vector $\boldsymbol{V}$. When (4.11) is used, we obtain

$$\kappa\{Q^{jk}L_{km} + \lambda^2 R^{-2}('\rho + 'p)\delta^j_m\}V^m \stackrel{*}{=} \lambda e^{-2\psi}R^{-4}\{\partial_0\partial_i\psi - (\dot{R}R^{-1} + \tfrac{1}{2}\partial_0\psi)\partial_i\psi\}L^{ij}, \tag{5.1}$$

and hence $\{V^m\}$ is uniquely determined provided $-\lambda^2('\rho + 'p)R^{-2}$ is not an eigenvalue of $\{Q^{jk}L_{km}\}$. We henceforth confine our attention to *nondegenerate* models; namely, those for which $-\lambda^2('\rho + 'p)R^{-2}$ is not an eigenvalue of $\{Q^{jk}L_{km}\}$, and hence the dispersion vector is uniquely determined. This determination is similar to that obtained from (4.17).

With Λ, ρ, p, R and $\{L_{ij}\}$ known from the corresponding classical model, the system (4.11) through (4.13), together with the trace-free condition on $\boldsymbol{Q}$ (see (4.7)) gives eleven independent equations for the determination of the thirteen unknowns $'\rho$, $'p$, $'\Lambda$, ψ, $\{V^i\}$ and $\{Q^{ij}\}$. We are thus at liberty to assign two additional conditions if a deterministic system is to result. At this point we shall actually add the three conditions

$$Q^{ij}V_j \stackrel{*}{=} 0 \tag{5.2}$$

in view of the simplicity thus afforded. These three conditions are not essential to the analysis since the nondegenerate models always yield a unique solution for the dispersion vector; their inclusion, however, leads to conformally homogeneous models that have many properties in common with the corresponding homogeneous models, as we shall now see.

By (4.6) and (4.7), the conditions (5.2) have the effect of yielding

$$Q^{00} \stackrel{*}{=} Q_{00} \stackrel{*}{=} Q^{0i} \stackrel{*}{=} Q_{0i} \stackrel{*}{=} 0, \quad Q^{ij} L_{ij} \stackrel{*}{=} 0. \tag{5.3}$$

The conditions (5.2) thus have the desired effect that the tensor $\boldsymbol{Q}$, which describes the state of anisotropy of the conformally homogeneous models, is a strictly space-like tensor in any comoving coordinate system of the associated homogeneous model. There is thus no time-wise anisotropy and no space-time anisotropic coupling in any comoving coordinate system when (5.2) holds. Since $\boldsymbol{U}$ is already known to be a null vector of $\boldsymbol{Q}$ (see (3.3)), (3.5) and (5.2) give

$$Q^{ij} W_j \stackrel{*}{=} 0. \tag{5.4}$$

A model that satisfies the conditions (5.2) will thus be said to be *normal* since such models have the velocity vector of the corresponding classical model in the null space of $\boldsymbol{Q}$ – a situation that tends to minimize the physical deviations between the two kinds of models.

When (5.2) is substituted into (5.1), we obtain

$$V^j \stackrel{*}{=} U^j \stackrel{*}{=} \frac{L^{ij}\{\partial_0 \partial_i \psi - (\dot{R} R^{-1} + \frac{1}{2}\partial_0 \psi)\partial_i \psi\}}{\kappa('\rho + 'p)\lambda R^2 e^{2\psi}} \tag{5.5}$$

for normal models. Since the inhomogeneous models are assumed to be nonempty, $'\rho + 'p \neq 0$, and accordingly the dispersion vector of a normal conformally homogeneous model vanishes only at those points of $\mathscr{C}$ for which

$$(\dot{R} R^{-1} + \tfrac{1}{2}\partial_0 \psi - \partial_0)\partial_i \psi \stackrel{*}{=} 0 \tag{5.6}$$

holds[18]. The implications of the normality condition are rather severe,

[18] It is of interest to examine the form which the conformal coefficient must take in order that the dispersion vector vanish throughout $\mathscr{C}$; that is, when (5.6) holds throughout $\mathscr{C}$. Under the substitution

$$u(x^K) = \exp(-\tfrac{1}{2}\psi(x^K)), \tag{1}$$

(5.6) becomes

$$2u^{-1}\partial_i(\partial_0 u - H(t)u) \stackrel{*}{=} 0, \quad H(t) \stackrel{*}{=} \dot{R}(t)R(t)^{-1}. \tag{2}$$

Since $u(x^K)$ is strictly positive for bounded ψ, we thus have the requirement

$$\partial_0 u - H(t)u \stackrel{*}{=} f(t) \tag{3}$$

where $f(t)$ is a C^1 function of t on $\mathscr{C}$. Since $u = u(x^K)$, the general solution of (3) is given by

$$u(x^K) \stackrel{*}{=} \exp(\textstyle\int H(t)dt)[k(x^i) + \int f(t)\exp(-\int H(t)dt)dt]. \tag{4}$$

The strict positivity of $u(x^K)$ on $\mathscr{C}$ requires that $k(x^i)$ and $f(t)$ be such that

$$k(x^i) + \textstyle\int f(t)\exp(-\int H(t)dt)dt \geq 0, \tag{5}$$

and accordingly $k(x^i)$ is any C^1 function on $\mathscr{C}$ such that (5) holds. We thus have

$$\psi \stackrel{*}{=} -2\ln\{\exp(\textstyle\int H(t)dt)[k(x^i) + \int f(t)\exp(-\int H(t)dt)dt]\} \tag{6}$$

for any C^1 functions $k(x^i)$ and $f(t)$ that satisfy (5) if the dispersion vector is to vanish throughout $\mathscr{C}$.

however. When (4.12) is contracted with $\{V^i\}$ and (4.13) is used, (5.2) yields

$$V^j\{\tfrac{1}{2}\partial_i\psi\,\partial_j\psi - V_j\partial_i\psi\} \overset{*}{=} [-\kappa\{('\rho + 'p)e^{\psi} - (\rho + p)\} \\ + 2V^2('\rho + 'p)e^{2\psi} + \dot{R}R^{-1}\partial_0\psi - \partial_0\partial_0\psi + \tfrac{1}{2}(\partial_0\psi)^2]\,V_i\,, \tag{5.7}$$

and hence normality obtains only under these conditions.

Now that $\{V^i\}$ has been determined for normal models, we can calculate V^2:

$$V^2 \overset{*}{=} R^2 L_{ij} V^i V^j \overset{*}{=} L^{ij} f_i f_j/(v^2 R^2 \lambda^2)\,,$$

where

$$f_i \overset{*}{=} \{\partial_0 - \dot{R}R^{-1} - \tfrac{1}{2}\partial_0\psi\}\partial_i\psi\,, \qquad v = \kappa('\rho + 'p)e^{2\psi}. \tag{5.8}$$

Hence, since $\lambda^2 = e^{-\psi} + V^2$, we obtain

$$V^2 \overset{*}{=} \tfrac{1}{2}e^{-\psi}\left(\sqrt{1 + 4v^{-2}R^{-2}L^{ij}f_i f_j e^{2\psi}} - 1\right) \tag{5.9}$$

and

$$\lambda \overset{*}{=} \sqrt{\tfrac{1}{2}e^{-\psi}}\sqrt{1 + \sqrt{1 + 4v^{-2}R^{-2}L^{ij}f_i f_j e^{2\psi}}}\,. \tag{5.10}$$

When these results are substituted into (5.5), we finally obtain

$$V^i \overset{*}{=} U^i \overset{*}{=} \frac{\sqrt{2}\,e^{\psi/2}L^{ij}f_j}{vR^2\sqrt{1 + \sqrt{1 + 4L^{km}f_k f_m v^{-2}R^{-2}}}} \tag{5.11}$$

6. Approximation: Prestructures

Up to this point the analysis has been exact. We now consider an approximation in which particularly simple situations arise in order to demonstrate the intrinsic utility of conformal homogeneous models and to show that such models yield structures characteristic of interesting and plausible inhomogeneities.

We consider those models for which the product of any two or more derivatives of ψ may be neglected in comparison with terms that are either linear in the derivatives of ψ or are independent of such derivatives. Such models correspond to cases in which the factor of proportionality between the line element of the classical models and the line element of the conformally homogeneous models has a wide range of possible values, but the change of the values of this factor in both space and time is sufficiently small that squares of such changes are negligible. An examination of the system of Eqs. (4.11) through (4.13) shows that the basic source of the differences between the classical models and the conformally homogeneous ones arise because of the derivatives of ψ. The linearizing approximation considered in this section may thus be viewed as yielding a first approximation to the actual inhomogeneities in the universe. We consequently refer to structures described by such a linear approximation as *prestructures.*

If $\sim$ denotes equality in a $*$-coordinate system to within the accuracy stated above, we have

$$\Box^2 \psi \sim \partial_0 \partial_0 \psi - R^{-2} \nabla^2 \psi + 3\dot{R}R^{-1}\partial_0 \psi, \quad \chi \sim 0, \tag{6.1}$$

where

$$\nabla^2 \psi = L^{ij}(\partial_i \partial_j \psi - \Gamma^k_{ij}(L)\partial_k \psi)$$

is the Laplacian of ψ evaluated in the 3-dimensional Riemann space $\mathscr{S}$, and

$$\begin{aligned} \lambda &\sim e^{-\frac{1}{2}\psi}, & f_i &\sim (\partial_0 - \dot{R}R^{-1})\partial_i \psi, \\ V^i &\sim e^{\frac{1}{2}\psi} L^{ij} f_j \nu^{-1} R^{-2}, & V^2 &\sim 0. \end{aligned} \tag{6.2}$$

When these results are substituted into (4.13) and the squares of the derivatives of ψ are neglected, we obtain

$$(\kappa\,'\rho + '\Lambda)e^{\psi} - (\kappa\rho + \Lambda) \sim 3\dot{R}R^{-1}\partial_0 \psi - R^{-2}\nabla^2 \psi, \tag{6.3}$$

while a combination of (6.3) and (4.14) yields

$$(\kappa\,'p - '\Lambda)e^{\psi} - (\kappa p - \Lambda) \sim -\partial_0 \partial_0 \psi - \tfrac{2}{3}(3\dot{R}R^{-1}\partial_0 \psi - R^{-2}\nabla^2 \psi). \tag{6.4}$$

It should be noted that (6.3) and (6.4) continue to hold if we relax the assumption of normality of the models but assume that $\{V^i\}$ is of the same order as a derivative of ψ. In this event, we again have $\lambda \sim e^{-\frac{1}{2}\psi}$ since $V^2 \sim 0$ continues to hold.

It was noted in Section 5, prior to the assumption of the normality condition, that we have eleven independent equations in thirteen unknowns, and hence we could add two conditions to the system. Since (6.3) and (6.4) hold without the assumption of the normality condition provided only that $V^2 \sim 0$, a situation that should certainly hold for prestructures, we are still free to make two assumptions. The interesting situation obtains in the case where we assume that $'\Lambda = 0$[19] and

$$\kappa\,'\rho \sim (\kappa\rho + \Lambda)e^{-\psi}\ ^{20}. \tag{6.5}$$

The assumption (6.5) gives $\exp(-\psi)$ a direct and simple physical interpretation. In this sense, $\exp(-\psi)$ becomes a simple physical descriptor of the inhomogeneities to be found in prestructures.

---

[19] Since the quantity $\kappa' p h_{AB}$ absorbs the trace of κQ, we may equally well assume that it absorbs the trace-like term $'\Lambda h_{AB}$. The assumption of the vanishing of $'\Lambda$ can thus be eliminated by a judicious reformulation and hence is a very mild constraint.

[20] The reason for $\kappa\rho + \Lambda$ appearing in this equation rather than just $\kappa\rho$ follows from the first of (1.6), where it is seen that $\kappa\rho + \Lambda$ is what is determined be the cosmological quantities k, $R(t)$ and $\dot{R}(t)$.

When the above assumptions are used in conjunction with (6.3), we obtain the following equation for the determination of ψ:

$$\nabla^2 \psi \sim 3 R \dot{R} \partial_0 \psi \,. \tag{6.6}$$

This equation will be referred to as the *prestructure equation.* Eq. (6.4) now gives us

$$\kappa' p \sim (\kappa p - \Lambda - \partial_0 \partial_0 \psi) \mathrm{e}^{-\psi} \,, \tag{6.7}$$

while (4.12) yields

$$R^4 \mathrm{e}^{2\psi} L_{ik} L_{jm} Q^{km} \sim -\partial_i \partial_j \psi + \Gamma^k_{ij}(L) \partial_k \psi + R \dot{R} L_{ij} \partial_0 \psi \,. \tag{6.8}$$

Accordingly, the normality condition is satisfied to within the approximation considered here, and (4.8), the trace-condition on $\boldsymbol{Q}$ also holds under $\sim$-equality (i.e. $L_{ij} Q^{ij} \sim 0$).

The relations (6.5), (6.7) and (6.8), together with (6.2), show that all quantities of interest are determined once ψ has been obtained by solving the prestructure Eq. (6.6). Our task thus devolves to that of obtaining C^2 solutions of the prestructure equation on the 3-dimensional space $\mathscr{S}$ of constant curvature k.

It is useful, at this point, to check the dimensional homogeneity of the prestructure Eq. (6.6). First, since ψ is a conformal coefficient, it is a dimensionless quantity. If L and T denote generic dimensions of length and time, respectively, then (1.1), (1.2) and (2.1), together with $[R] = L$, yield $[L_{ij}] = T^2 L^{-4}$, where $[\mathrm{d} x^i] = L$. Accordingly, $[\nabla^2 \psi] = [L^{ij} \nabla_i \nabla_j \psi] = T^{-2} L^2$, and $[R \dot{R} \partial_0 \psi] = T^{-2} L^2$.

7. The Condition of Bounded Potential in the Classical Models

The problem of describing prestructures has been shown to be equivalent to the problem of solving the prestructure Eq. (6.6). Accordingly, since (6.6) involves the quantities k, $R(t)$ and $R(t)$ of the classical models, we digress with a more detailed discussion of the classical models so as to determine k and $R(t)$. As is well known, the classical models with the standard observables admit many possible universes. We therefore restrict our considerations by the introduction of what appears to be a new observable within the context of cosmology [21].

The existence of an upper bound for the local value of the gravitational potential that may obtain anywhere in the universe is suggested by both theory and observation. Following Schwarzschild, several

[21] Edelen, D. G. B., Wilson, A. G.: *Astrophys. J. 151,* 1171 (1968).

authors[22] have shown theoretically that the potential $\Phi = 2Gm/(c^2 r)$ (G = gravitational coupling constant, c = speed of light, and m = total mass contained within a sphere of radius r) of a static, spherically symmetric system immersed in a region of zero mass density is bounded. Examples of some of these theoretical potential limits and their delineating assumptions are given in Table 1. In addition to the bounds suggested by theory, there is also observational evidence suggesting the existence of a potential bound for stable nondegenerate cosmic bodies[23].

Table 1. *Relativistic potential bounds*

Bound	Symbol	Constraints: Spherical symmetry plus	Upper limit to $2Gm/c^2 r$
Schwarzschild	U_S	$\rho \equiv$ Const.	1
Eddington	U_E	$\rho \equiv$ Const., p finite	0.888
Bondi I	U_B	ρ not increasing from center, $p \leq \rho c^2/3$	0.638[a]
Bondi II	U_A	Adiabatic stability, $p \leq \rho c^2/3$	0.620[a]

[a] Instead of the proper radius, Bondi uses $\sqrt{(A/4\pi)}$, where A is the proper area.

Table 2. *Maximum observed gravitational potentials*

System	Object	$\log_{10}(m/r)$ (g/cm)
Star	V 444 (Cyg A)	23.27
Galaxy	M 87	23.6
Cluster	Coma	23.5

The maximum values of the mass/radius ratios that occur in samples of stars, galaxies and clusters of galaxies that have been measured are found to be nearly the same for each species of cosmic body. Table 2 gives the largest values of the observed potentials[24]. The same values may also be derived for second-order clusters on the basis of their radii and cluster counts[25]. Although these facts present themselves, the value, U_0, of this observed potential bound and its significence are uncertain.

[22] Eddington, A. S.: *The Mathematical Theory of Relativity*. Cambridge: Cambridge University Press 1923; Tolman, R. C.: *Relativity, Thermodynamics and Cosmology*. Oxford: Clarendon Press 1934; Buchdahl, H. A.: *Phys. Rev. 116*, 1027 (1959); Chandrasekhar, S.: *Astrophys. J. 140*, 417 (1964); Bondi, H.: *Proc. Roy. Soc. Ser. A 282*, 303 (1964); Fowler, W. A.: *Trans. I. A. U. 12 B*, 581 (1966).

[23] Wilson, A. G.: *Astron. J. 71*, 402 (1966).

[24] Allen, C. W.: *Astronomical Quantities*. 2nd ed. London: Athlone Press 1963.

[25] Vaucouleurs, G. de: *Soviet Astron. AJ 3*, 897 (1960); Abell, G.O.: *Astron. J. 66*, 607 (1961).

The observational and theoretical results stated above suggest the following hypothesis: *there exists a global potential upper bound, U, such that, for any sphere of sufficiently large proper radius, r, circumscribed about any point P as center, the total mass contained within the sphere will satisfy the inequality*

$$\Phi = 2Gm/(c^2 r) \leq U \,. \tag{7.1}$$

The bounded potential condition, (7.1), will now be applied to the classical models with the usual assumption of simply connected covering spaces. We will not take the value of the speed of light to be unity in this section in order to simplify numerical comparisons, and hence the line element is given by

$$\mathrm{d}s(g)^2 \overset{*}{=} c^2 \mathrm{d}t^2 - R(t)^2 \mathrm{d}u(L)^2 \,, \tag{7.2}$$

where we now use the equivalent form

$$\mathrm{d}u(L)^2 \overset{*}{=} \mathrm{d}l^2 + \varepsilon(l)^2 (\mathrm{d}\theta^2 + \sin^2\theta \,\mathrm{d}\varphi^2) \tag{7.3}$$

with

$$\varepsilon(l) = \begin{cases} \sin(l), & 0 \leq l \leq \pi \,, \quad \text{for} \quad k = 1 \,, \\ l, & 0 \leq l \leq \infty \,, \quad \text{for} \quad k = 0 \,, \\ \sinh(l), & 0 \leq l \leq \infty \,, \quad \text{for} \quad k = -1 \,. \end{cases}$$

The field Eqs. (1.6) are then given by

$$\kappa \rho c^2 \overset{*}{=} -\Lambda + 3(k + \dot{R}^2 c^{-2}) R^{-2} \,, \tag{7.4}$$

$$\kappa p \overset{*}{=} \Lambda - 2\ddot{R} c^{-2} R^{-1} - (k + \dot{R}^2 c^{-2}) R^{-2} \,. \tag{7.5}$$

If $m(X,t)$ denotes the mass within the "coordinate sphere" $0 \leq l \leq X$ at time t, then, since the mean mass density, ρ, is a function of time alone, we have

$$m(X,t) = \rho R^3 \int_0^X \varepsilon(l)^2 \mathrm{d}l \int_0^\pi \sin\theta \,\mathrm{d}\theta \int_0^{2\pi} \mathrm{d}\varphi \,.$$

As l is a geodesic coordinate, the proper (geodesic) radius, $r(X,t)$, of the "comoving coordinate sphere" $0 \leq l \leq X$ at time t will be

$$r(X,t) = R \int_0^X \mathrm{d}l = RX \,.$$

The potential on the surface $l = X$ is thus given by

$$\begin{aligned} \Phi &= 2Gm(X,t)/(c^2 r(X,t)) \\ &= 8\pi G \rho R^2 c^{-2} \begin{cases} \frac{1}{2}(1 - \sin(2X)/(2X)), & \text{for} \quad k = 1 \,, \\ X^2/3, & \text{for} \quad k = 0 \,, \\ \frac{1}{2}(\sinh(2X)/(2X) - 1) \,, & \text{for} \quad k = -1 \,. \end{cases} \end{aligned} \tag{7.6}$$

In the cases of zero and negative curvatures ($k = 0$ and $k = -1$), the allowable range of X is infinite. Since $R(t_0) = R_0 \neq 0$ at the present epoch (the subscript 0 is used to designate values at the present epoch), it follows that the inequality (7.1) can be satisfied for all $X > 0$ only if $\rho = 0$. This leads to the conclusion that, for *any* finite potential bound, the only universes with zero and negative curvatures that are isotropic and homogeneous have zero mean mass density. Furthermore, this conclusion holds regardless of the dynamical processes involved, since it was obtained without use of the field equations. Note should be taken that this conclusion may not be valid if the spaces are not simply connected. There exist eighteen different topological forms for a 3-dimensional space with zero curvature and infinitely many in the case of negative curvature. Some of these spaces are known to be closed, in which event the argument given above ceases to hold.

A similar argument can not be made in the case $k = 1$, since the function

$$S(X) = 1 - \sin(2X)/(2X)$$

occurring in (7.6) is bounded for all X. Accordingly, in order to discuss the implications for bounded potentials in this case, we must use the field equations. For $k = 1$, we consider only the two customary topological cases: *spherical space* for which $0 \leq X \leq \pi$, and *elliptical space* for which $0 \leq X \leq \pi/2$. At the values $X = \pi$ and $X = \pi/2$, the so-called spherical and elliptical horizons, respectively, the potential Φ takes on the same value, namely $\kappa c^2 \rho R^2/2$. In what follows it will be convenient to introduce the quantity

$$\Phi(t)^* = \kappa c^2 \rho R^2 \tag{7.7}$$

equal to twice the horizon value of the potential. By (7.6) and (7.7), together with the definition of $S(X)$, the inequality (7.1) becomes

$$\Phi^* S(X)/2 \leq U. \tag{7.8}$$

In spherical space, $S(X)$ assumes the maximal value of 1.217 ... at $X = 2.245$ ... radians, and hence

$$\Phi^* \leq 1.644\, U. \tag{7.9}$$

In elliptical space, $S(X)$ assumes the maximal value of unity at the "end point" $X = \pi/2$, and hence

$$\Phi^* \leq 2U. \tag{7.10}$$

In order to make use of these bounds on Φ^*, we write the field Eqs. (7.4) and (7.5) for $k = 1$ in terms of Φ^* and its time derivative:

$$\Phi^* - 3 = 3\dot{R}^2 c^{-2} - \Lambda R^2, \tag{7.11}$$

$$\Phi^* + \dot{\Phi}^*/H = -3\kappa p R^2, \tag{7.12}$$

where $H = \dot{R}/R = H(t)$ is Hubble's parameter at time t. For any reasonable model, the pressure is bounded from below by zero and from above by the photon gas pressure $\rho c^2/3$. We accordingly introduce the dimensionless parameter n with values on the closed interval $[0, 1]$[26] and write

$$p = n\rho c^2/3. \tag{7.13}$$

The field Eq. (7.12) then reads

$$\dot{\Phi}^* = -(1+n)\Phi^* H. \tag{7.14}$$

With the additional definitions

$$\sigma = 4\pi G\rho/(3H) = c^2 \Phi^* \dot{R}^{-2}/6,$$
$$q = -R\dot{R}^{-2}\ddot{R},$$

where q is the "deceleration parameter", we easily derive

$$q = (1+n)\sigma - \Lambda c^2/(3H^2) \tag{7.15}$$

$$-\dot{\Phi}^*/(2H) = \Lambda R^2 + 3\dot{R}^2 q c^{-2}. \tag{7.16}$$

For all potential bounds less than or equal to the Schwarzschild limit ($U_S = 1$), the left member of (7.11) is strictly negative in some neighborhood of the present epoch. The cosmological constant Λ must accordingly be greater than zero. With the upper and lower bounds on the pressure ($n = 1$ and $n = 0$), the definition of Φ^*, and (7.14), we obtain the inequalities

$$0 \leq \Phi^*/2 \leq -\dot{\Phi}^*/(2H) \leq \Phi^*. \tag{7.17}$$

Substitution of the right-hand inequality into (7.16) and adding (7.11), we have

$$(1+q)\dot{R}^2 c^{-2} \leq 2\Phi^*/3 - 1. \tag{7.18}$$

This inequality demands that $q < -1$ whenever $\Phi^* < 3/2$.

At this point, it is useful to point out the relation between the value of the deceleration parameter obtained here and the value, q_f, of the deceleration parameter used in the Friedmann models as reported by Sandage[27]. In the Friedmann models, $n = \Lambda = 0$, and hence (7.15) gives $q_f = \sigma$. Accordingly, by (7.15), we may write

$$q = (1+n)q_f - \Lambda c^2/(3H^2). \tag{7.19}$$

We shall show later that the system of equations used here implies that

$$\Lambda = \kappa c^2 \rho(3-\Phi^*)/\Phi^* + 3H^2 c^{-2}, \tag{7.20}$$

and accordingly, for $\Phi^* < 2$, we have

$$\Lambda \geq \tfrac{1}{2}\kappa c^2 \rho + 3H^2 c^{-2} = c^{-2}(4\pi G\rho + 3H^2). \tag{7.21}$$

[26] In general, n is a function of ρ and possibly of other physical parameters such as temperature, etc. For our purposes, however, it is sufficient to consider n as a simple parameter that serves to differentiate between various world models.

[27] Sandage, A. R.: *Astrophys. J. 133*, 355 (1961).

A substitution of this bound on Λ into (7.19) then gives

$$q_0 \leq (1+n_0)q_{f0} - 1 - \kappa c^2 \rho_0/(6H_0^2). \tag{7.22}$$

Currently estimated values of q_{f_0}, derived from densities and ages, indicate small positive values in the neighborhood of about 0.02. With n_0 negligible, we thus obtain $q_0 < -0.98 - \kappa c^2 \rho_0/(6H_0^2)$, and hence our predicted value is consistent with observations to within the accuracy that can be attributed to them.

Summerizing the above considerations, we have shown that Λ *must be positive* for bounded potentials and nonvanishing mass, $k = 1$, the deceleration parameter is negative, and (7.21) together with $H_0 =$ 75 km/mps, $\rho_0 = 3.1 \cdot 10^{-31}$ gm/cm^3, give $\Lambda \geq 2 \cdot 10^{-56}$/cm^2.

8. The Future State with Bounded Potential

We may now investigate the allowable evolutionary paths of classical models with bounded potentials by using the conclusions of the previous Section as initial conditions for Eqs. (7.11) and (7.14). In this way we can determine the functions $R(t)$ and $\dot{R}(t)$, and hence ρ and p.

Since $\dot{\Phi}_0^* < 0$ and $\Phi_0^* > 0$ (for U_S or any smaller bound), there will exist some interval T containing the present epoch, t_0, during which the left member of (7.11) will remain strictly negative. If we define the function $b(t)$ by the relation

$$b(t) = \sqrt{(3-\Phi^*)/\Lambda}, \tag{8.1}$$

the relations

$$3\dot{R}^2/(\Lambda c^2) - R^2 = -b(t)^2 < 0 \tag{8.2}$$

hold for all t on the interval T and $b(t)$ is real valued on T. The change of variables given by

$$R = R(t) = b(t) \cosh w(t) \tag{8.3}$$

is thus permitted, and the equality in (8.2) becomes

$$\{(\dot{b}/b)\cosh w + \dot{w} \sinh w\}^2 = (c^2 \Lambda/3)\sinh^2 w. \tag{8.4}$$

Accordingly, (8.3) will be a solution of the equality in (8.2) provided $w(t)$ satisfies the equation

$$\dot{w} = c\sqrt{\Lambda/3} - (\dot{b}/b)\coth w. \tag{8.5}$$

The positive square root has been selected in obtaining (8.5) and this selection obviously entails no loss of generality since the sign of the initial value of $w(t)$ is at our disposal. We consequently must have $w_0 > 0$, in view of (8.7) and $\dot{R}_0 > 0$. Differentiating (8.1), we get

$$\dot{b} = b\dot{\Phi}^*/(2\Phi^* - 6), \tag{8.6}$$

while (8.3) gives

$$\dot{R} = c\sqrt{\Lambda/3}\; b \sinh w \tag{8.7}$$

and hence

$$H = c\sqrt{\Lambda/3} \tanh w\,. \tag{8.8}$$

When the above results are combined in the obvious manner and (7.14) is used, we obtain the governing equations

$$\dot{\Phi}^* = -A(1+n)\Phi^* \tanh w \tag{8.9}$$

and

$$\dot{w} = 2A - A(1+n)\Phi^*/(3-\Phi^*), \tag{8.10}$$

where $A = \frac{1}{2}c\sqrt{\Lambda/3}$. An elimination of Λ between (8.1) and (8.7) and the definition of σ give us

$$\sinh^2 w = \Phi^*/(2\sigma(3-\Phi^*)),$$

and hence an elimination of the hyperbolic functions between this result and (8.3) gives (7.20). If we treat the parameter n as a constant, we may then derive

$$\ddot{w} = 6A^2(1+n)^2 \tanh w\; \Phi^*/(3-\Phi^*), \tag{8.11}$$

$$\ddot{\Phi}^* = -2A^2(1+n)\Phi^*\{(2-(1+n)\Phi^*/(3-\Phi^*))\operatorname{sech}^2 w - 2(1+n)\tanh^2 w\}\,. \tag{8.12}$$

We can now read off the future history. Since $\dot{\Phi}^*_0 < 0$ and $(\Phi^*_0 - 3) < 0$, we have $b_0 > 0$ and $b(t)$ is an increasing function of t by (8.6). Eq. (8.3) then shows that the full behavior of $R(t)$ is controlled essentially by the function $w(t)$. By (8.10), we have $\dot{w} > 0$ whenever

$$\Phi^* < 6/(3+n)\,. \tag{8.13}$$

At the present epoch, n is negligible and hence $w(t)$ is an increasing function of t. Accordingly, both $R(t)$ and $\dot{R}(t)$ are essentially exponentially increasing into the future. These results continue to hold for all future time since Φ^* is a decreasing function by (8.9) (recall that $w_0 > 0$). The various asymptotic values can then be read off from the above relations whenever the condition (8.13) holds.

It is customary to assume that the cosmological equations continue to hold into the indefinite or distant past, whatever the case may be, even when they predict singular states or states quite different from the present state. In this regard, the inequality (8.13) establishes a line of demarcation, for satisfaction of (8.13) always results in exponential behavior. If this inequality is rigorously satisfied for all past time, the universe has an essentially hyperbolic cosine behavior (sans the function $b(t)$), as shown by (8.3). On the other hand, if (8.13) is not satisfied after some time in the past[28], a number of alternatives may develop among which is the possibility of $w(t)$ oscillating and $R(t)$ oscillating along the cosh curve as a consequence of changes in n. Numerous other problems associated with an adequate description of the genesis process are discussed by Harrison[29].

9. Solution of the Prestructure Equation and the Resulting Inhomogeneities

The condition of bounded potential for the classical models provides us with the necessary information for solving the prestructure equation

$$\nabla^2 \psi = 3 R \dot{R} \partial_0 \psi , \tag{9.1}$$

for we know that, for $\rho \neq 0$, we must have $k = 1$ while $R(t)$ behaves essentially like the hyperbolic cosine and $\dot{R}(t)$ behaves essentially like the hyperbolic sine in the neighborhood of the present epoch and in the future[30]. Accordingly, if we take an initial time for which $\dot{R} > 0$ and such that the inequality (8.13) is satisfied, then (9.1) represents a regular diffusion process with a time dependent diffusion coefficient. A separation of variables gives us

$$\psi = \sum_j X_j(x^k)\, T_j(t)$$

where

$$T_j(t) = A_j \exp\left\{ -\frac{\lambda_j}{3} \int_{t_0}^{t} R(y)^{-1} \dot{R}(y)^{-1} \,\mathrm{d}y \right\} \tag{9.2}$$

and

$$\nabla^2 X_j = -\lambda_j X_j . \tag{9.3}$$

Now, ∇^2 is the Laplacian on the 3-dimensional space $\mathscr{S}$ and, since $k = 1$, this space is a compact, simply connected and complete Riemann space that is isomorphic to the surface of the unit sphere in Euclidean 4-dimensional space[31]. By the theorem of Bochner[32], (9.3) admits nontrivial, C^2 solutions if and only if the values of λ_j are positive and discrete; in which case, the solutions of (9.3) (eigenfunctions) are convex functions about the trivial solution $X_j = 0$, and we have

$$\lambda_j = j(j+2) \tag{9.4}$$

---

[28] This possibility must be considered since (8.9) shows that Φ^* increases into the past so long as $w(t) > 0$.

[29] Harrison, E. R.: *Physics Today 21*, 31 (1968).

[30] Recall that the argument of these hyperbolic functions is $w(t)$ and that $w(t)$ is an increasing function in the neighborhood of the present epoch and in the future with $w_0 > 0$.

[31] Edelen, D. G. B.: *J. Math. Anal. Appl. 23*, 99 (1968).

[32] Bochner, S.: *Duke Math. J. 3*, 334 (1937).

for positive integer values of j[33]. The full set of eigenfunctions forms a complete system of orthogonal functions on the space $\mathscr{S}$ and defines scales of inhomogeneity; each scale being defined by the diameter of the cells into which 3-space is divided by the surfaces $X_j(x^k) = 0$. Now, the scale of an inhomogeneity will decrease monotonically with increasing λ (varying approximately as $\lambda_j^{-\frac{1}{2}} \sim (j+1)^{-1}$) and (9.2) shows that the relaxation time associated with each such scale decreases as λ_j increases. The large-scale inhomogeneities will thus decay more slowly than the small-scale ones. The possibility is thus presented for describing large-scale inhomogeneities, such as galaxies, clusters of galaxies and second order clusters which are such that they have not had time to dissipate during the present expansion phase of the universe. An adequate accounting for small-scale inhomogeneities, on the other hand, will require the full nonlinear equations without the prestructure approximation since the condition $\|\partial_A\psi\,\partial_B\psi\| \ll \|\partial_A\psi\|$ will not be satisfied. In this respect, it is even problematic as to whether a discussion of acceptable rigour can be made for galaxies on the basis of the prestructure approximation. In any event, the prestructure results indicate that conformally homogeneous models provide an acceptable basis for the description of the enveloping environment of galaxies, and, as such, gives what we need for our considerations in the next Chapter.

For the prestructure approximation, the above solutions give

$$\begin{aligned} f_i &\sim (\partial_0 - \dot{R}R^{-1})\partial_i\psi = \sum T_j(H + \lambda_j/(3R^2H))\partial_i X_j, \\ V_i &\sim e^{\frac{1}{2}\psi} L^{ij} f_j v^{-1} R^{-2}, \end{aligned} \tag{9.5}$$

and hence the points where the dispersion vector vanishes are those where $X(x^k)$ possesses a local maximum or minimum. Accordingly, since $'\rho/\rho \sim e^{-\psi}$, $'\rho$ is maximum when $X(x^k)$ is minimum. It thus follows that $\boldsymbol{U} \sim \lambda\boldsymbol{W}$ at a cluster center, supercluster center, etc. Dispersion thus always exists for the conformally homogeneous prestructure models, vanishes only at points of spatial maximum or minimum of the density distribution, and decays in time both with ψ and with $R(t)^{-2}$. Further, from the above $\sim$-equality of $\{V^i\}$, the dispersion vector points toward the centers of mass concentration. Mass concentrations may thus be viewed as dynamically bound entities whose binding decays with time as the universe expands.

One further point should be noted here. Since ψ is of the form $\sum_j X_j(x^k)T_j(t)$, while $\rho/\rho_0 \sim e^{-\psi}$, we accordingly have

$$\rho/\rho_0 \sim \prod_j \exp(-X_j(x^k)T_j(t)). \tag{9.6}$$

[33] See Edelen, D. G. B.: *J. Math. Anal. Appl.* *23*, 99 (1968), for explicit forms for the eigenfunctions and for the calculations of the eigenvalues.

It thus follows that the contributions from the various modes and the corresponding time-relaxations combine multiplicatively rather than additatively. Further, (9.2) shows that the eigenvalues that show up in the $T_j(t)$'s occur in an exponent, and hence the general term in the product (9.6) is of the form

$$\exp\left\{-X_j(x^k)A_j \exp\left[-\frac{j(j+2)}{3}\int_{t_0}^{t} R(y)^{-1}\dot{R}(y)^{-1}\,dy\right]\right\}.$$

B. Internal Consistency Structure: The Locally Induced Environment

Because the resolving power of any optical or radio telescope has a definite lower bound, the fundamental observables in distant galaxies derive from aggregate properties of stars, gases, and dust clouds rather than from the individual properties of these constituents. Thus, if theory and observation are to mesh, the models of galactic structures should be representative of aggregate or averaged behavior; indeed, it is only through considerations of such averages that mass density and luminosity are meaningfully associated with galaxies. As the domains of general relativity and classical relativistic cosmology are currently envisioned, galaxies and clusters of galaxies play the role of indicator points whose properties must themselves be averaged in order to determine the salient parameters of the various models. Unfortunately, classical cosmology denies the very existence of such indicator points and hence a description of the averaging processes and domains of averaging that are consistent with the classical models can not be obtained within the discipline of classical cosmology.

On the other hand, the Einstein field equations are experimentally known to relate only the local differential geometric quantities (curvature and infinitesimal metric structure) and the local momentum-energy distribution that is characteristic of a stellar neighborhood. Since these field equations and their associated physical interpretations are carried over directly as a basis for the classical cosmological models and for the description of galaxies and clusters of galaxies, a fundamental problem must be resolved – namely, the construction of the governing equations of aggregates when the individual elements of the aggregates are governed by the Einstein field equations. A solution of this problem will not only give a clear understanding of the gross structures of the universe in terms of the known properties of their constituents, it will also define the possible structural environment for galaxies that is demanded by the conditions of internal consistency. Knowledge of this internal consistency environment is obviously necessary if we are to come to grips with the problems involved in obtaining an acceptable description of galactic structures.

10. Observation Operators

First, we need a conceptual description of the observation process that is reasonably precise and that is also compatible with the geometric and physical structures under examination in the Einstein theory.

We start with a 4-dimensional metric space $\mathscr{E}$ of the hyperbolic-normal type, whence the signature of the metric tensor of $\mathscr{E}$ is -2. Let $\boldsymbol{O}$ denote a generic point of $\mathscr{E}$ which we identify with the position of the observer at the instant of observation. Such an observer erects a tetrad of linearly independent contravariant vectors $\{W_{\mathfrak{a}}^{A}(\boldsymbol{O})\}$ at $\boldsymbol{O}$ by means of which he can reduce any linear homogeneous geometric object (vectors, tensors, etc.) at the point $\boldsymbol{O}$ to an ordered array of scalars by the process of transvection[34]. Thus, if $\{T_A(\boldsymbol{O})\}$ is a vector field at $\boldsymbol{O}$, the corresponding scalars are given by

$$T_{\mathfrak{a}}(\boldsymbol{O}) = T_A(\boldsymbol{O}) W_{\mathfrak{a}}^{A}(\boldsymbol{O}) . \tag{10.1}$$

Physically, the scalars $\{T_{\mathfrak{a}}(\boldsymbol{O})\}$ may be viewed as the values of the field $\{T_A(\boldsymbol{O})\}$ when this field is resolved on the "laboratory" or "reference" frame defined by the tetrad $\{W_{\mathfrak{a}}^{A}(\boldsymbol{O})\}$.

Commencing with the tetrad $\{W_{\mathfrak{a}}^{A}(\boldsymbol{O})\}$, we can extend it to a field of tetrads $\{W_{\mathfrak{a}}^{A}(\boldsymbol{x})\}$ over $\mathscr{E}$ in any number of different ways. For instance, $\{W_{\mathfrak{a}}^{A}(\boldsymbol{x})\}$ can be defined by parallel transport of $\{W_{\mathfrak{a}}^{A}(\boldsymbol{O})\}$ along the geodesic in $\mathscr{E}$ that connect the points $\boldsymbol{O}$ and $\boldsymbol{x}$, provided $\boldsymbol{x}$ lies in the neighborhood of $\boldsymbol{O}$ for which the geodesic connectivity is single valued. A possibly more useful method is to take $\{W_{\mathfrak{a}}^{A}(\boldsymbol{O})\}$ as a linearly independent system of (two complex conjugate and two real) null vectors at $\boldsymbol{O}$ and then to imbed $\{W_{\mathfrak{a}}^{A}(\boldsymbol{O})\}$ in a normed null basis on $\mathscr{E}$[35]. If this done, the tetrad forms what may be referred to as a "radiation frame" since two of its constituent vectors are real null vectors that are tangent to the null cone at each point of $\mathscr{E}$. Such a structure is certainly a natural one in dealing with astronomical bodies since the radiation frame is tailor-made for representing information that is propagated over the null cone (light and radio signals).

Regardless of the manner in which $\{W_{\mathfrak{a}}^{A}(\boldsymbol{O})\}$ is extended to a system of tetrad fields on $\mathscr{E}$, other than the obvious smoothness conditions and the requirement

$$\det(W_{\mathfrak{a}}^{A}(\boldsymbol{x})) \neq 0 \tag{10.2}$$

---

[34] The reader is reminded of the conventions concerning indices; in particular, that lower case German letters are used to denote labeling indices and are not tensor indices.

[35] Edelen, D. G. B.: *J. Math. Mech. 16*, 351 (1966); – *J. Math. Mech. 16*, 435 (1966); – *J. Math. Mech. 16*, 885 (1967).

for all $\boldsymbol{x}$ in $\mathscr{E}$, a reference frame is defined at each point of $\mathscr{E}$ by the tetrad field. In view of the requirement (10.2), there exists a unique inverse (dual) system of fields $\{W_A^{\mathfrak{a}}(\boldsymbol{x})\}$ defined by

$$W_A^{\mathfrak{a}}(\boldsymbol{x})\, W_{\mathfrak{b}}^{\ A}(\boldsymbol{x}) = \delta_{\mathfrak{b}}^{\mathfrak{a}}\,, \qquad W_A^{\mathfrak{a}}(\boldsymbol{x})\, W_{\mathfrak{a}}^{\ B}(\boldsymbol{x}) = \delta_A^B\,.\qquad (10.3)$$

[36]

If the fields $\{W_A^{\mathfrak{a}}(\boldsymbol{x})\}$ happen to be such that

$$\nabla_{[A}\, W_{B]}^{\mathfrak{a}} = 0\,, \qquad (10.4)$$

the covariant tetrad is said to form a holonomic frame[37]. If (10.4) is not satisfied, the tetrad is said to form an anholonomic frame[38]. It is probably evident to the reader that most frames of reference used in physics are anholonomic frames since they are chosen primarily for the convenience of the experimentalist rather than for their natural adaptability to an underlying geometry in the global sense (satisfaction of (10.4) throughout $\mathscr{E}$). We consequently assume that the frame defined by the tetrad under consideration is anholonomic since this assumption results in no loss of generality.

Any linear geometric object field at any point of $\mathscr{E}$ can be reduced to an ordered array of scalars by transvection with the appropriate members of the anholonomic frame and its dual at the point in question. For example,

$$t_{\mathfrak{a}\mathfrak{b}}^{\mathfrak{c}}(\boldsymbol{x}) = W_C^{\mathfrak{c}}(\boldsymbol{x})\, W_{\mathfrak{a}}^{\ A}(\boldsymbol{x})\, W_{\mathfrak{b}}^{\ B}(\boldsymbol{x})\, T_{AB}^{C}(\boldsymbol{x})\,.$$

Such collections of scalars represent the result that would be obtained by an observer at the point $\boldsymbol{x}$ in $\mathscr{E}$ by the resolution of the field $\{T_{AB}^{C}(\boldsymbol{x})\}$ with respect to the reference frame defined by $\{W_{\mathfrak{a}}^{\ A}(\boldsymbol{x})\}$. Mathematically, the scalars $\{t_{\mathfrak{a}\mathfrak{b}}^{\mathfrak{c}}(\boldsymbol{x})\}$ are referred to as the anholonomic components of the field $\{T_{AB}^{C}(\boldsymbol{x})\}$ (with respect to $W_{\mathfrak{a}}^{\ A}$) and depend in an intrinsic fashion on the choice of the reference frame. It is accordingly obvious that the

[36] These relations motivate the reference to a tetrad and its dual as the intermediary components of the identity; Hlavatý, V.: *The Geometry of Einstein's Unified Field Theory*. Groningen: Noordhoff 1957.

[37] Schouten: Sect. 9, p. 99ff.

[38] The term "anholonomic coordinates" is prevalent in the literature and arose principally due to the foibles of history. It is really the frame that is holonomic or anholonomic and this has nothing to do with coordinates or coordinate coverings; anholonomic coordinates are almost a contradiction in terms since they give a pseudo-coordinatization of a manifold that is path dependent and arises from pathwise integration of nonintegrable Pfaffians, as opposed to natural coordinatizations that result from the integration of completely integrable Pfaffians. Anholonomic coordinates arose because of the convenience afforded in interpreting an anholonomic frame of reference in terms of the differential of a coordinate transformation at a point. The problem with anholonomic coordinates comes in the attempt to extend the coordinate differential interpretation to a finite neighborhood of a point.

relations between anholonomic components of fields at different points involve not only the changes in the field values from point to point, but also on the manner in which the anholonomic reference frame varies from point to point. This dependence on the anholonomic reference frame may be viewed as the inseparable dependence of observed quantities on the processes and techniques of observation.

Let us identify ourselves with the observer at $\boldsymbol{O}$, and ask what we may infer by observation as to the state of affairs at a generic point $\boldsymbol{x}$ of $\mathscr{E}$. For simplicity, let us assume that we are concerned with a covariant vector field $\{T_A(\boldsymbol{x})\}$. If $\boldsymbol{y}$ denotes a generic point of $\mathscr{E}$, an observer at $\boldsymbol{y}$ would record, with respect to his reference frame $\{W_{\mathfrak{a}}^{A}(\boldsymbol{y})\}$, the scalars

$$t_{\mathfrak{a}}(\boldsymbol{y}) = W_{\mathfrak{a}}^{A}(\boldsymbol{y})\, T_A(\boldsymbol{y})\,. \tag{10.5}$$

If there were no error in the observations and if there were a perfect communication process between $\boldsymbol{y}$ and $\boldsymbol{O}$, the observer at $\boldsymbol{O}$ could claim that the field of scalars he would "see" at $\boldsymbol{x}$ would be given by $\{t_{\mathfrak{a}}(\boldsymbol{x})\}$[39]. It then follows, on the assumption that the observer at $\boldsymbol{O}$ knows the dual frame $\{W_B^{\mathfrak{b}}(\boldsymbol{y})\}$, that a transvection (on German indices) of (10.5) with $\{W_B^{\mathfrak{b}}(\boldsymbol{y})\}$ reproduces $\{T_B(\boldsymbol{y})\}$.

In actual fact, any observation process always involves certain intrinsic errors and imperfect communication processes. In addition, it is quite often necessary to restrict the investigation to certain ensemble properties or averages due to the inability of the observer to control or affect the observation process or its environment. Under these circumstances, the field of scalars that an observer at $\boldsymbol{O}$ ascribes to a point $\boldsymbol{x}$ would be a composite of contributions from field values at points neighboring to $\boldsymbol{x}$. This has the effect of converting a point-valued field into a set-valued field.

In order to account for the errors in the observation process and its intrinsic limitations, we must introduce some additional notation. Let $M_{\mathfrak{c}}^{\mathfrak{b}}(\boldsymbol{y},\boldsymbol{x};\boldsymbol{O})$ denote the contribution to the $\mathfrak{c}^{\text{th}}$ scalar that an observer at $\boldsymbol{O}$ would ascribe to the point $\boldsymbol{x}$ due to the unit $\mathfrak{b}^{\text{th}}$ scalar at the point $\boldsymbol{y}$ per unit of invariant (volume) measure $dv(\boldsymbol{y})$ of $\mathscr{E}$ at the point $\boldsymbol{y}$[40]. It then follows that the composite scalar fields that an observer at $\boldsymbol{O}$ would ascribe to the point $\boldsymbol{x}$ under these circumstances would be given by

$$t_{\mathfrak{c}}(\boldsymbol{x};\boldsymbol{O}) = \int t_{\mathfrak{b}}(\boldsymbol{y})\, M_{\mathfrak{c}}^{\mathfrak{b}}(\boldsymbol{y},\boldsymbol{x};\boldsymbol{O})\, dv(\boldsymbol{y})\,. \tag{10.6}$$

[39] In effect, this amounts to the assumption that the observer at $\boldsymbol{O}$ can follow his reference frame as it evolved on passage from $\boldsymbol{O}$ to $\boldsymbol{y}$. This ability is what is hidden in the assumption that there is a perfect communication process between $\boldsymbol{y}$ and $\boldsymbol{O}$.

[40] In most cases, it is sufficient to take $dv(\boldsymbol{y})$ to be the proper infinitesimal volume element of $\mathscr{E}$ at $\boldsymbol{y}$ since any singularities or the observation process may be accounted for by an appropriate restriction of the support of the scalars $M(\boldsymbol{y},\boldsymbol{x};\boldsymbol{O})$ or by the extension of these quantities to generalized scalar functions in the sense of distributions.

When this is transvected (on German indices) with $\{W_B^{\mathfrak{c}}(x)\}$, we obtain the vector field

$$\langle T_B(x)\rangle = W_B^{\mathfrak{c}}(x)\int t_{\mathfrak{b}}(y)M_{\mathfrak{c}}^{\mathfrak{b}}(y,x;O)\,\mathrm{d}v(y) \tag{10.7}$$

that an observer at O would ascribe to the point x. That $\langle T_B(x)\rangle$ always constitutes the components of a vector field follows from previously published results[41].

If we combine the above equations, we obtain

$$\langle T_B(x)\rangle = \int \mathscr{K}_B^A(y,x;O)\,T_A(y)\,\mathrm{d}v(y) \tag{10.8}$$

where

$$\mathscr{K}_B^A(y,x;O) = W_{\mathfrak{a}}^A(y)M_{\mathfrak{b}}^{\mathfrak{a}}(y,x;O)\,W_B^{\mathfrak{b}}(x). \tag{10.9}$$

On the other hand, if we had started with a contravariant vector field $\{T^A(y)\}$, then (10.6) would have been replaced by

$$t^{\mathfrak{b}}(x;O) = \int t^{\mathfrak{a}}(y)N_{\mathfrak{a}}^{\mathfrak{b}}(y,x;O)\,\mathrm{d}v(y), \tag{10.10}$$

where $N(y,x;O)$ plays the same role for contravariant quantities as $M(y,x;O)$ plays for covariant quantities. Accordingly, we obtain

$$\langle T^B(x)\rangle = \int \mathscr{L}_A^B(y,x;O)\,T^A(y)\,\mathrm{d}v(y) \tag{10.11}$$

with

$$\mathscr{L}_A^B(y,x;O) = W_A^{\mathfrak{a}}(y)N_{\mathfrak{a}}^{\mathfrak{b}}(y,x;O)\,W_{\mathfrak{b}}^B(x). \tag{10.12}$$

The operator $\langle\cdot\rangle$ is referred to as the *observation operator*. The extension of this operator to tensors and tensor densities of arbitrary order is immediate from the multilinear structure of such quantities:

$$\langle T_{A\cdots}^{B\cdots}(x)\rangle = \int \mathscr{K}_A^C\cdots\mathscr{L}_D^B\cdots T_{C\cdots}^{D\cdots}(y)\,\mathrm{d}v(y). \tag{10.13}$$

The natural duality between covariant and contravariant quantities usually rubs off on the M's and the N's, for in most instances it turns out that the application of the observation operator to the identity tensor, $\{\delta_A^B\}$, gives a nonzero multiple of this tensor. In this way, the algebraic structure of tensor fields is almost preserved and significant eliminations of pathology of a purely mathematical nature is afforded. With (10.13) and the preceeding relations, we have

$$\langle\lambda(x)\delta_A^B\rangle = \int \mathscr{K}_A^C(y,x;O)\mathscr{L}_D^B(y,x;O)\lambda(y)\delta_C^D\,\mathrm{d}v(y),$$

and hence

$$\langle\lambda(x)\delta_A^B\rangle = \langle\lambda(x)\rangle\,\delta_A^B \tag{10.14}$$

if and only if

$$\mathscr{K}_A^C(y,x;O)\mathscr{L}_C^B(y,x;O) = \mathscr{S}(y,x;O)\delta_A^B \tag{10.15}$$

$$\langle\lambda(x)\rangle = \int \mathscr{S}(y,x;O)\lambda(y)\,\mathrm{d}v(y). \tag{10.16}$$

---

[41] Edelen, D. G. B.: *J. Math. Mech. 13*, 927 (1964).

A simple translation of (10.15) in terms of the M's and the N's gives

$$M_{\mathfrak{a}}^{\mathfrak{b}}(y,x;O)N_{\mathfrak{b}}^{\mathfrak{c}}(y,x;O)=\mathscr{S}(y,x;O)\delta_{\mathfrak{a}}^{\mathfrak{c}}, \tag{10.17}$$

while (10.16) defines the effect of the observation operator on a scalar function. We shall assume the validity of the relations (10.14) through (10.17), which may be viewed as a "normalization that preserves the identity". Completeness and regularity then demands that

$$\int \mathscr{S}(y,x;O)\,\mathrm{d}v(y)>0 \tag{10.18}$$

for all x in $\mathscr{E}$.

We have consistently retained the implicit dependence of these results on the observer by the inclusion of the argument O. This is no quirk of nicety or happenstance – rather, it entails a fundamental belief that the observer can not be removed from the picture until there is specific data that demands it.

11. From the Local to the Macroscopic

We denote the *local* momentum-energy tensor by $\{t_{AB}\}$ and the corresponding local metric tensor by $\{g_{AB}\}$. By a local tensor field we mean the tensor field an observer would obtain from the world structure in the immediate neighborhood of his star system. Consequently, in the first approximation obtained by ignoring the contributions to the momentum-energy tensor that obtain from the planets, comets, solar wind, solar radiation, etc., we would take the Schwarzschild exterior solution for the local metric tensor. Since the Einstein field equations are known to hold for our immediate stellar neighborhood, the g's are solutions of the equivalent systems of field equations

$$R_{AB}(g)-\tfrac{1}{2}R(g)g_{AB}=\kappa t_{AB}, \tag{11.1a}$$

$$R_{A}^{B}(g)-\tfrac{1}{2}R(g)\delta_{A}^{B}=\kappa t_{A}^{B}, \tag{11.1b}$$

$$R^{AB}(g)-\tfrac{1}{2}R(g)g^{AB}=\kappa t^{AB}, \tag{11.1c}$$

where $\{R_{AB}(g)\}$ and $R(g)$ are the Ricci tensor and the scalar curvature formed from the metric tensor $\{g_{AB}\}$.

We can neither observe nor work directly with the local tensor fields in both galactic research and cosmological research. What is done, in effect, is to replace the local fields by certain averaged or ensemble fields in much the same fashion that continuum mechanics is obtained from point mechanics – although the analogy is closer in the case of statistical mechanics and point mechanics. It is a fact, however, that for relativistic formulations, the process of averaging or of constructing ensembles remains to be given a precise and unequivocal formulation[42].

We take as a basic tenent the validity of the Einstein field Eqs. (11.1) for the description of a stellar neighborhood. If we apply the observation operator to the three equivalent systems of field Eqs. (11.1), we obtain three systems of "aggregate field equations" that are not equivalent. This inequivalence arises because the observation operator does not commute, in general, with the raising or lowering of tensor indices by use of the metric tensor and its inverse. In view of the assumptions stated by (10.14) through (10.17), the evident choice is to base the analysis on the equations that result from application of the observation operator to the system (11.1 b). This choice effectively places the burden equally on the covariant structure and on the contravariant structure, and the duality of these two structures leads to a formulation that exhibits a minimal deviation from the locally valid field equations[43]. We thus obtain

$$\langle R_A^B(g)\rangle - \tfrac{1}{2}\langle R(g)\rangle \delta_A^B = \kappa \langle t_A^B\rangle \tag{11.2}$$

as the resulting aggregate field equations.

Now, $\{\langle t_A^B\rangle\}$ is the averaged or macroscopic momentum-energy tensor, and hence the inferable physical properties and symmetries of the aggregate would be obtained from this tensor. In fact, it is this tensor field that is assigned by physical intuition or by detailed considerations of ensembles such as those given by Synge[44]. On the other hand, if we set

$$h_{AB}(x) = \langle g_{AB}(x)\rangle\,, \tag{11.3}$$

so that $\{h_{AB}\}$ is the macroscopic or aggregate metric tensor, it is easily seen that $\{\langle R_A^B(g)\rangle\}$ is generally not the Ricci tensor formed from either $\{g_{AB}\}$ or from $\{h_{AB}\}$. A little reflection on the construction and properties of the observation operator, and the fact that the Einstein field equations are nonlinear, show that the system of field Eq. (11.2) governing the aggregate have different properties than the Einstein field equations themselves. In particular, it is no longer true that the covariant divergence of $\{\langle t_A^B\rangle\}$, formed from the connections based on either $\{g_{AB}\}$ or

[42] See Havas, P.: *International Symposium on Statistical Mechanics and Thermodynamics*. Technische Hochschule, Aachen 1964, for a penetrating survey of some of the associated problems in relativistic statistical mechanics.

[43] See Shirokov, M. F., Fisher, I. Z.: *Astron. Ž. 39.* 899 (1962) (*Soviet Astron. A J. 6,* 699 (1963)), for a detailed physical argument that leads to the choice of (11.1)b. Care must be exercised in applying the Shirokov and Fisher results, for they use an averaging operator that does not preserve the tensor character of the descriptors – the integral of a tensor is not a tensor in general.

[44] Synge, J. L.: *Relativity: The General Theory,* p. 159ff. Amsterdam: North-Holland 1960.

on $\{h_{AB}\}$, vanish. This follows since the contracted Bianchi identities and the Bianchi identities[45] themselves do not hold for $\{\langle R^B_A(g)\rangle\}$. It thus follows that *the aggregate* Eqs. (11.2) *do not by themselves lead to equations of motion for the aggregate.*

In general, there are three possible interpretations of the above situation. The first, which we do not pursue here, is that the assumption of metrizability can not be maintained in the description of aggregate phenomena – the affine connection of the space may have to be assumed to be more general than that of a metric space, such as in unified field theory[46] or in variant field theory[47]. The second is that an entirely different approach must be taken to obtain the field equations for aggregates, such as that used in wave geometry[48]. The third is that the metric assumption and the approach adopted here is correct, but that $\{\langle t^B_A\rangle\}$ does not adequately describe the physical interplay of the aggregate. The latter interpretation would seem to be the more likely situation, since it is well known that information is always lost in the process of averaging and the success in the use of the Einstein field equations in cosmology. In this respect, if we define the *commutator tensor*, $\{C^B_A\}$, by the relations

$$C^B_A = R^B_A(h) - \langle R^B_A(g)\rangle + \tfrac{1}{2}(\langle R(g)\rangle - R(h))\delta^B_A\,, \tag{11.4}$$

the h's satisfy what may be referred to as the *macroscopic Einstein field equations*

$$R^B_A(h) - \tfrac{1}{2}R(h)\delta^B_A = \kappa T^B_A \tag{11.5}$$

provided the *macroscopic momentum-energy tensor*, $\{T^B_A\}$, is defined by

$$T^B_A = \langle t^B_A\rangle + \kappa^{-1} C^B_A\,. \tag{11.6}$$

The resulting field Eqs. (11.5) are then identical in formal structure with the Einstein field equations and the field equations of cosmology, with the exception that the metric tensor is now played by the tensor $\{h_{AB}\}$. In view of (11.3), we consequently refer to $\{h_{AB}\}$ as the *macroscopic metric tensor.*

Due to the occurrence of the tensor $\{C^B_A\}$, which will not, in general, be a null tensor, the macroscopic momentum-energy tensor $\{T^B_A\}$ will

---

[45] Schouten: p. 146ff.

[46] Hlavaty, V.: *The Geometry of Einstein's Unified Field Theory*. Groningen: Noordhoff 1957.

[47] Edelen, D. G. B.: *The Structure of Field Space*. Berkeley: University of California Press 1964; – *J. Math. Mech. 13*, 927 (1964).

[48] Mimura, Y., Takeno, H.: *Wave Geometry* (Scientific Reports of the Research Institute for Theoretical Physics, Hiroshima Univ., No. 2, 1962); Mimura, Y., Takeno, H., Sakuma, K., Ueno, Y.: *Wave Geometry II* (Scientific Reports of the Research Institute for Theoretical Physics, Hiroshima Univ., No. 6, 1967).

not exhibit the same physical symmetries or properties as evidenced by the observable aggregate momentum-energy tensor $\{\langle t_A^B \rangle\}$. What this says is that the constituent granular structure of an aggregate is reflected in its macroscopic representation by the presence of deviations from the geometric and physical properties expected from the observable averages of the local momentum-energy complexes. From this it is natural to interpret the quantities $\{\kappa^{-1} C_A^B\}$ as the increments of macroscopic momentum-energy that are required in order to equilibrate the fluctuations present in the local geometric and momentum-energy complexes. Further, from (11.5), (11.6) and the Bianchi identities on the metric space $\mathscr{E}^*$ with metric tensor $\{h_{AB}\}$, one immediately obtains the *equations of motion for the aggregate:*

$$\nabla_B \langle t_A^B \rangle + \kappa^{-1} \nabla_B C_A^B = 0 \,. \tag{11.7}$$

The covariant differentiation occurring in (11.7) is formed from the Christoffel symbols of the second kind based on the metric tensor $\{h_{AB}\}$ of the space $\mathscr{E}^*$, and this macroscopic metric space is, in general, distinct from the metric space $\mathscr{E}$. Eqs. (11.7) show that one must include the presence of the *effective force vector* $\kappa^{-1}\{\nabla_B C_A^B\}$, which arises as a consequence of the granular nature of the matter that makes up the observed aggregate. This force vector accounts for the information that is lost because we are only able to observe aggregate properties while the actual motions of the constituent elements of the aggregate are determined by the individual interactions of the constituents.

12. Implication and Inference

We noted in the previous Section that it is the tensor $\{\langle t_A^B \rangle\}$, not $\{T_A^B\}$, that is assigned for aggregates by observation or by calculation. Thus, if we accept the working hypothesis that the Einstein field equations hold in our stellar neighborhood, the usual arguments used in cosmology[49] or in galactic research lead to the determination

$$\langle t_A^B \rangle = (\rho + p) W_A W^B - p \delta_A^0, W_A = h_{AB} W^B, W^A W_B = 1 \,, \tag{12.1}$$

where ρ, p and W^A are the mean density, mean pressure, and mean velocity field of the aggregate, respectively. A combination of (11.6) and (12.1) thus gives the following determination of the macroscopic momentum-energy tensor:

$$T_A^B = (\rho + p) W_A W^B - p \delta_A^B + \kappa^{-1} C_A^B \,. \tag{12.2}$$

The construction of the macroscopic environment from the local structure thus leads to an augmentation of the "perfect fluid" momentum-

---

[49] Robertson, H. P.: *Rev. Modern Phys.* 5, 62 (1933).

energy tensor that is similar in many respects to that obtained previously from the standpoint of representing the actual inhomogeneities from the enveloping cosmological viewpoint. Thus similarity of the augmentation of the momentum-energy structure, from both the small and from the large, will be of fundamental importance in the next Chapter. It can not be stressed too strongly that the physical arguments used in the past to determine the whole of the momentum-energy tensor for classical cosmological models, can, in actuality, only determine $\{T_A^B - \kappa^{-1} C_A^B\}$ if the Einstein field equations are to hold locally. On the other hand, if the results of the previous Section are examined in detail, it can be seen that if $C_A^B \neq 0$ for some pair of indices (A,B), setting $T_A^B = \langle t_A^B \rangle$ would imply that the Einstein field equations do not hold locally. We may consequently infer the following: if $C_A^B \neq 0$ and the Einstein field equations hold locally, the physical conclusions obtained from the classical cosmological models based on $T_A^B = \langle t_A^B \rangle$ must be reexamined in terms of the correct relations $T_A^B = \langle t_A^B \rangle + \kappa^{-1} C_A^B$.

An examination of the macroscopic field equations from the standpoint of determinism also leads to some surprises. When (11.5) and (11.6) are combined, we obtain

$$R_A^B(h) - \tfrac{1}{2} R(h) \delta_A^B = \kappa \langle t_A^B \rangle + C_A^B . \tag{12.3}$$

We therefore have a system of ten partial differential equations of the second order that are connected by four differential identities (the contracted Bianchi identities). In the general case, the number of unknowns is thirty – the ten independent h's, the ten independent $\langle t \rangle$'s, and the ten independent C's – in contrast to the situation that obtains in classical cosmology where the number of unknowns is twenty – the ten independent h's and the ten independent $\langle t \rangle$'s. If $\{\langle t_A^B \rangle\}$ is assumed to be given by (12.1), as is the usual case, the number of independent quantities involved in the $\langle t \rangle$'s is five – the three independent W's, ρ and p. In this case we would have a total of twenty five dependent variables that are connected by ten equations that must satisfy four differential identities. *If* we know the local tensors $\{g_{AB}\}$ and $\{t_A^B\}$ throughout $\mathscr{E}$, we could calculate the h's and the C's so that the macroscopic equations would have the same degree of determinism as the local equations. The g's and the t's are not known throughout $\mathscr{E}$, however, since only the h's and the $\langle t \rangle$'s are amenable to observation at distant points. Accordingly, the C's must be taken as unknowns for problems involving aggregates and in classical cosmology. Thus, if the Einstein field equations hold locally, the number of unknowns in a general relativistic description of aggregate structures is ten greater than entertained in previous theories. In this context, classical cosmology may be viewed as obtained under the ten constraint equations $C_A^B = 0$, and these constraint equations may be attributed as the statement of the cosmological principle within the context of the Einstein theory.

Although the C's are in general unknown in aggregate problems, they must satisfy the four constraint equations

$$\nabla_B \langle t_A^B \rangle = -\kappa^{-1} \nabla_B C_A^B . \tag{12.4}$$

Since $\langle t_A^B \rangle$ is observable and $\{h_{AB}\}$ is the macroscopic metric tensor, observation of the motions of the aggregate would lead to a determination of a vector with the components $\nabla_B \langle t_A^B \rangle \stackrel{\text{def}}{=} Q_A$. Accordingly, the C's must satisfy the system of linear equations

$$\nabla_B C_A^B = -\kappa Q_A . \tag{12.5}$$

The general solution of (12.5) is comprised of a particular solution of (12.5) and a general solution of the homogeneous equations

$$\nabla_B Y_A^B = 0 . \tag{12.6}$$

A combination of this result with (12.4) shows that $\{\langle t_A^B \rangle\}$ together with the observable motions of the aggregate can determine the macroscopic metric structure and the macroscopic momentum-energy structure only to within solutions of the homogeneous Eq. (12.6). It thus follows that the indeterminism in aggregate structures and in cosmology is real and can not just be wished away.

The macroscopic metric tensor $\{h_{AB}\}$ was introduced in the previous Section by the relations $h_{AB} = \langle g_{AB} \rangle$, while the averaging operator $\langle \cdot \rangle$ is basically defined in terms of its action on the local observables $\{t_A^B\}$ (the metric tensor is not itself a primitive observable). It is easily seen, however, that any symmetric, nonsingular tensor could have been introduced in place of $\{h_{AB}\}$ since the local g's are unobservables and are uninferable at distant points, and the effects of such an introduction could be compensated for by the appropriate choice of the C's. Further still, we may simply assign the h's as per the Weyl postulates as is done in classical cosmology. In this event, Eqs. (12.3) then serve to determine the C's provided the $\langle t \rangle$'s are assumed known. Ultimately, what this boils down to is the fact that the h's are artificial quantities that define a metric structure on an abstract, hyperbolic-normal, four-dimensional metric space $\mathscr{E}^*$ that need have no relation to the space time $\mathscr{E}$ of physical events in the local Einstein theory. As an immediate consequence of these observations, we obtain the fundamental conclusions listed below.

1. The h's may not be identified necessarily with the gravitational potentials as in the local Einstein theory.

2. The line integral of $(h_{AB} \mathrm{d}x^A \mathrm{d}x^B)^{\frac{1}{2}}$ along the geodesic in $\mathscr{E}^*$ that connects points P and Q does not determine the physical space-time interval between P and Q.

3. Since light travels along null geodesics defined by the local metric tensor $\{g_{AB}\}$, it will not travel along null geodesics defined by $\{h_{AB}\}$ unless $\{h_{AB}\}$ and $\{g_{AB}\}$ are conformally related; the latter situation being highly unlikely.

The last conclusion requires a reinvestigation of the meaning and significance of most cosmological observables previously interpreted in terms of classical cosmology, since light does not necessarily traverse null geodesics of the classical cosmological models. The compact galaxies of Zwicky [50] may thus be of first order significance due to their gravitational lense effects. It is also evident from the findings of Schmidt [51], Sandage [52] and Hewish *et al.* [53] of objects with red shifts between one and 2.012, a large population of quiet radio objects with ultraviolet excess, and rapidly pulsating radio sources, respectively, that something is awry in classical cosmology theory.

---

[50] Zwicky, F.: Discussion. Commission 28, XII I. A. U. (Hamburg, 1964).
[51] Schmidt, M.: *Astrophys. J. 141*, 1560 (1965).
[52] Sandage, A.: *Astrophys. J. 141*, 1295 (1965).
[53] Hewish, A. *et al.*: *Nature 217*, 709 (1968).

CHAPTER II

An Equilibrium Theory of Galactic Isopleths

In the original classification system of galaxies given by Hubble[1], the optical morphology of elliptical galaxies was described in terms of a single ellipticity parameter,

$$E = 10\,(\text{major axis} - \text{minor axis})/(\text{major axis})\,,$$

which ranged from zero through approximately seven for the observed sample. The upper limit of ellipticity, together with the ages of these galaxies (population II stars) could be interpreted readily in terms of equilibrium bodies of rotation. Although Hubble[2] anticipated a hypothetical transition class, *SO*, of galaxies between the ellipticals (*E* galaxies) and the spiral galaxies, with ellipticities greater than or equal to seven, these were not found until the photographic survey of the nearby galaxies (1936 – 50).

Detailed studies of a number of late ellipticals[3] and *SO*'s reveal that their contours frequently depart from true ellipsoids of revolution over a wide range of luminous densities. Although several subclasses of perturbed forms are readily identifiable[4], the wide variety of forms assumed by the contours and the fact that the isophotes frequently reverse their trend of change with respect to the luminous density greatly complicate the attempt to establish a morphological classification system based on empirical geometric parameters[5]. Further, it is difficult to

---

[1] Hubble, E.: *Astrophys. J. 64,* 321 (1926).

[2] Hubble, E.: *The Realm of the Nebulae.* New Haven: Yale University Press 1936.

[3] The designation, early ellipticals, refers to the elliptical galaxies with ellipticity near zero, while, late ellipticals, refers to elliptical galaxies with ellipticities between approximately five and seven.

[4] Vaucouleurs, G. de: Classification and Morphology of External Galaxies. In: *Handbuch der Physik*, vol. LIII. Berlin-Göttingen-Heidelberg: Springer 1959.

[5] Basic descriptive data concerning the types and forms encountered in the study of galactic morphology is given by Vaucouleurs, G. de: Classification and Morphology of External Galaxies. In: *Handbuch der Physik*, vol. LIII. Berlin-Göttingen-Heidelberg: Springer 1959; Sandage, A.: *The Hubble Atlas of Galaxies.* Washington: Carnegie Institution 1961, to mention just two. Particular note should be taken of Hodge, P. W.: *Astron. Soc. Pacific*, Leaflet No. 435 (1965), where a reasonably full discussion of *SO* galaxies is given.

understand such perturbed objects as equilibrium configurations in classical terms without the introduction of a significant number of tenuous assumptions.

Eventually, galactic morphology must be explained by functional relations between the observed geometric parameters and the physical parameters that govern the structure and dynamics of galaxies. Thus, since the primary task is to relate the geometric and dynamic parameters, it seems most straightforward to apply the Einstein field equations, for these equations directly relate the salient geometric and physical descriptors (tensors). The use of the Einstein field equations, however, demands a self-consistent and dynamically complete specification of the momentum-energy tensor of a galaxy, and this requires data that are not available at the present time. Further, the consideration of aggregates and the aggregate field equations, as given in Part B of the previous Chapter, must come into play since a galaxy certainly requires a description in terms of field equations that apply to aggregates; otherwise, the basic observables of galactic structure, such as luminous density and mass density fail to have any basis for their description within the theory. The unknown macroscopic momentum-energy tensor and the unknown commutator tensor of a galaxy thus make it impossible for us to discuss the internal dynamics of galaxies with any acceptable degree of confidence at the present time.

On the other hand, the basic observables pertaining to the structure of galaxies are sets of isophotol contours. This suggests that we direct our attention to sets of physical isoplethic surfaces analogous to the isophotes. Ideally, these isopleths would be surfaces of constant mean mass density or of constant mean luminous density, or the projection of these surfaces onto the plane of the sky. It thus transpires, in view of our lack of information concerning the dynamics, that if we are to have any success in obtaining properties of galactic isopleths under the Einstein theory, we must find some way of getting hold of the isoplethic surfaces in a fashion that does not involve a direct solution of the Einstein field equations or of their aggregate modification. Further, we should require the isoplethic surfaces to be characteristic of configurations in an appropriately defined state of equilibrium, for it is only under such circumstances that we may account for the ages of galaxies and preserve a smooth transition from the rotational equilibrium of the early ellipticals.

An equilibrium theory of galactic isopleths is developed in this Chapter. In many respects, the theory constitutes what has been referred to as a radical departure from past considerations. It relies on a strictly geometric definition of equilibrium isoplethic surfaces and obtains most of its results from a system of conditions that must be imposed if there are to exist solutions of the Einstein field equations in the presence of jump discontinuities in the momentum-energy tensor. These conditions and their analysis entail a somewhat protracted sequence of arguments, particularly in view of the lack of knowledge of the momentum-energy tensor with which we have to deal. However, the results thus obtained admit of many areas of applicability and provide us with the necessary basis for a theory of galactic morphology and scale.

A. Geometric Equilibrium

Most galaxies appear to have definite geometric configurations that persist for a time comparable to that during which a star of solar mass remains on the main sequence in the course of its natural evolution. If we were able to deal directly with the dynamics of galaxies, such a situation would connote an equilibrium process. The facts, however, refer to geometric configurations from which very little dynamical information can be gleaned, both because of the problems associated with the description of aggregates and because of the tremendous linear scales and distances of galactic structures and their highly diffuse natures[6].

This part is devoted to the development of an alternative concept of equilibrium whereby a smooth transition from the ellipticals to the *SO*'s may be effected and in which all of the geometric shapes in the sequence define equilibrium constructs. This concept will be developed within the context of general relativity, for it is within this context that there exists a suitable structure for acceptable relations between geometric equilibrium and physical equilibrium.

13. Isopleths and Timelike Hypersurfaces

If we have a diffuse dynamical object such as a galaxy, we can study certain aspects of its structure by investigating the evolution of the isoplethic surfaces, such as surfaces of constant mean mass density or of constant mean luminous density. In the space-time, $\mathscr{E}$, of general relativity, these isoplethic surfaces together with the time variable lead to three-dimensional timelike hypersurfaces that are the world tubes of the isopleths.

We assume that coordinate systems have been chosen in $\mathscr{E}$ such that the line element $ds^2 = h_{AB}(x^K)dx^A dx^B$ takes the form

$$ds^2 = V(x^K)^2(dx^0)^2 - g_{ij}(x^K)dx^i dx^j, \tag{13.1}$$

[6] Mean mass densities in the neighborhood of 10^{-24} gm/cm^3.

where $\{g_{ij}(x^K)\}$ is positive definite and $V(x^K) \neq 0$ throughout $\mathscr{E}$. This assumption places no restriction on the generality of the analysis, for Thomas[7] has shown that such coordinate systems always exist under the usual continuity assumptions. In the context of the Einstein theory, we may consider x^0 as a generalized time variable and the $\{x^i\}$ as coordinates of a one-parameter family of three-dimensional Riemann spaces $\mathscr{R}(t)$ with the fundamental metric differential forms

$$\mathrm{d}L(t)^2 = g_{ij}(x^K)\big|_{x^0=t}\mathrm{d}x^i\mathrm{d}x^j. \tag{13.2}$$

Alternatively, we may consider $\mathscr{R}(t)$ to be the image of $\mathscr{R}(0)$ under transport along the x^0-axis. Under this interpretation, the $\{x^i\}$ become convected coordinates.

Let $\mathscr{S}$ denote a regular, timelike hypersurface in $\mathscr{E}$ with the implicit and parametric representations

$$f(x^K) = 0 \quad \text{and} \quad x^A = f^A(u^\Gamma)\,^8.$$

Here, the quantities $\{u^\Gamma\}$ denote the parametric or surface coordinates on $\mathscr{S}$. If we refer $\mathscr{E}$ to a coordinate system in which (13.1) holds, the intersection, $\mathscr{S}(t)$, of the hypersurface $\mathscr{S}$ with the spacelike hypersurface $x^0 = t$ lies in the Riemann space $\mathscr{R}(t)$, and is what we normally think of as the configuration of an ordinary two-dimensional surface in three-dimensional space at the generalized time t. Consequently, if we are to identify $\mathscr{S}$ with the world tube of an isoplethic surface of a galaxy, we must require $\mathscr{S}(t)$ to be a *closed* two-dimensional surface in the Riemann space $\mathscr{R}(t)$ for all t in the interval of interest.

There is associated with any regular hypersurface $\mathscr{S}$ two symmetric tensor fields $\{a_{\Gamma\Sigma}(u^\Lambda)\}$ and $\{b_{\Gamma\Sigma}(u^\Lambda)\}$, known as the coefficient tensors of the first and second fundamental forms[9], and a vector field $\{N^A(f^B)\}$ of normals (outwardly oriented) to the hypersurface in the enveloping $\mathscr{E}$. If $\{x^A_\Gamma\}$ denote the projectors $\{\partial f^A(u^\Lambda)/\partial u^\Gamma\}$, we have

$$a_{\Gamma\Sigma} = h_{AB}(f^K)x^A_\Gamma x^B_\Sigma, \qquad a^{\Gamma\Sigma}x^A_\Gamma x^B_\Sigma = h^{AB}(f^K) + N^A N^B, \tag{13.3}$$

the tensor $\{a^{\Gamma\Sigma}\}$ being used to denote the inverse of $\{a_{\Gamma\Sigma}\}$. Since $\mathscr{S}$ is a regular, timelike hypersurface, the rank of the matrix $((x^A_\Gamma))$ is three and the signature of $\{a_{\Gamma\Sigma}\}$ is -1. The quantities $\{a_{\Gamma\Sigma}\}$ also serve to determine the metric structure on $\mathscr{S}$ by the quadratic form

$$\mathrm{d}l^2 = a_{\Gamma\Sigma}(u^\Lambda)\mathrm{d}u^\Gamma\mathrm{d}u^\Sigma. \tag{13.4}$$

---

[7] Thomas, T. Y.: *Tensor 14*, 199 (1963).

[8] In accordance with common usage, a hypersurface in $\mathscr{E}$ is timelike if and only if its normal is spacelike; that is

$$\partial_A f\,\partial_B f\,h^{AB}(f^K) < 0$$

at all points of $\mathscr{S}$.

[9] Schouten: p. 242ff.

We consider an observer attached to the hypersurface $\mathscr{S}$ and constrained in such a way that he can only make measurements in this hypersurface, while the direction of the normal to $\mathscr{S}$ in the enveloping space $\mathscr{E}$ is unknown to him. The space and time measurements of such an observer would be determined by $\mathrm{d}l^2$ and he would consequently conclude that he lived in a three-dimensional metric space $\mathscr{S}^*$ whose line element is given by (13.4). In view of the signature of $\{a_{\Gamma\Sigma}\}$, the three-dimensional space $\mathscr{S}^*$ is the hyperbolic-normal metric space of the intrinsic geometry of $\mathscr{S}$ that would be recorded by such a hypothetical observer.

Since $\mathscr{S}^*$ is the three-dimensional space-time of the observers that are confined to $\mathscr{S}$, there exists a timelike vector field on $\mathscr{S}^*$ with components $Y^\Gamma(u^\Lambda)$ that defines a proper time orientation for the observers. We accordingly have

$$Y^\Gamma Y^\Sigma a_{\Gamma\Sigma} = \exp(-2\Psi(u^\Lambda)) > 0 \tag{13.5}$$

throughout $\mathscr{S}^*$. The congruence of curves, $\mathscr{K}$, given on $\mathscr{S}^*$ by solutions of the equations $\mathrm{d}u^\Gamma/\mathrm{d}p = Y^\Gamma(u^\Lambda)$, defines a time-oriented congruence, each curve of which defines the motion of a clock that ticks the proper time of the coincident observer. The curves $\mathscr{K}$ are thus proper time lines of $\mathscr{S}^*$.

We now need to establish some form of calibration and synchronization of the clocks of the observers moving along the proper time lines, $\mathscr{K}$, of $\mathscr{S}^*$. To this end, let us refer the four-dimensional space-time $\mathscr{E}$ to a coordinatization in which (13.1) holds. We have seen that the intersection of $\mathscr{S}$ with the hypersurface $x^0 = t$ in $\mathscr{E}$ defines the physical two-dimensional isopleth $\mathscr{S}(t)$ in $\mathscr{R}(t)$, and hence we may calibrate the clocks on the time lines of $\mathscr{S}^*$ at a time $\bar{t}$ by requiring them to have the same reading on the corresponding $\mathscr{S}^*(\bar{t})$ in $\mathscr{S}^*$. As time proceeds from the calibration time $\bar{t}$, the infinitesimal two-dimensional space elements of observers traveling along the proper time congruence will synchronize if these local two-dimensional isoplethic surfaces mesh at every later time t into the corresponding two-dimensional isoplethic surface in $\mathscr{R}(t)$. This simply states that synchronized observers on $\mathscr{S}^*$ stay with the physical two-dimensional isopleth as time proceeds; the observers ride through time with the two-dimensional isopleth. Since the proper time congruence is defined in terms of the vector field $\{Y^\Gamma\}$, one would expect that the requirement of synchronization could be expressed in terms of this vector field. This is indeed the case, as a necessary and sufficient condition for such synchronization is given by

$$Y_{[\Gamma}\nabla_\Sigma Y_{\Lambda]} = 0, \tag{13.6}$$

where the indicated (surface) covariant differentiation is formed from the affine connection of the metric space $\mathscr{S}^*$. An alternative description of this requirement is that $\{Y^\Gamma\}$ forms a Fermi-irrotational vector field [10] on $\mathscr{S}^*$ – the proper time lines of the observers do not twist around the world tube $\mathscr{S}$. In a more prosaic vein, all of the observers grasp hands immediately after calibration and hold each other as rigidly as possible so that they do not rotate relative to each other; the best excuse for an inertial frame when there is nothing of an intrinsic nature to glue such a frame to. A vector field that satisfies the conditions (13.6) is such that the congruence $\mathscr{K}$ is a *normal congruence* and $\{Y^\Gamma\}$ is perpendicular to the two-dimensional physical isopleth with respect to the measure of angle inferred from the metric structure of $\mathscr{S}^*$.

14. Geometric Equilibrium

Having glued the observers to the two-dimensional isopleth and having synchronized them so that they stay with the isopleth as the time varies, we can begin to consider the question of equilibrium. As each observer moves along the world tube of the isopleth as the time varies, he can make observations as to how his neighbors move with respect to him and how his clock varies with respect to his neighbors'; he can observe how $\mathrm{d}l^2$ varies from point to point along his proper time line. In view of (13.4), this is equivalent to observing how $\{a_{\Gamma\Sigma}\}$ changes along his proper time line. Now, the change in a tensor field that is measured by an observer moving along a curve with tangent vector $\{Y^\Gamma\}$ is given by the Lie derivative of the tensor field with respect to the vector field $\{Y^\Gamma\}$ [11]. In the case of the metric tensor $\{a_{\Gamma\Sigma}\}$ of $\mathscr{S}^*$, the Lie derivative with respect to $\{Y^\Gamma\}$ has the evaluation $2\nabla_{(\Sigma} Y_{\Gamma)} = \nabla_\Sigma Y_\Gamma + \nabla_\Gamma Y_\Sigma$. Since knowledge of $\{\nabla_{(\Gamma} Y_{\Sigma)}\}$ determines how the isopleth changes with respect to time in terms of the observer riding with the isopleth, a specification of this quantity controls the dynamics of the isopleth [12].

An isopleth with a system of observers that satisfy the conditions (13.5) and (13.6) would be representative of an equilibrium configuration if there were no change in the spatial separations of the observers in the course of time. The formulation is in a hyperbolic-normal, three-dimensional space $\mathscr{S}^*$, however, and hence this requirement is that the projection of $\{\nabla_{(\Gamma} Y_{\Sigma)}\}$ normal to $\{Y^\Gamma\}$ should vanish [13]. However, if no further

[10] Edelen, D. G. B.: *Proc. Nat. Acad. Sci. U.S.A, 49*, 598 (1963).

[11] Schouten: p. 102ff.; Yano, K.: *Theory of Lie Derivatives*. Amsterdam: North-Holland 1957.

[12] Specification of the Lie derivative of the metric tensor with respect to a given vector field that forms a normal congruence contains a significent amount of information. In the case of the Einstein theory, the momentum-energy tensor can be determined from this information: Edelen, D. G. B.: *Arch. Rational Mech. Anal. 16*, 316 (1964).

[13] Recall that $\{Y^\Gamma\}$ defines the proper time orientation of the observers and hence projection normal to $\{Y^\Gamma\}$ yields the intrinsic space-like part.

conditions are assumed, the proper time measures would be quite complicated functions of the coordinate separations[14], and hence an essential feature of what is normally meant by equilibrium would be lost; namely, the object does not change with respect to a universal time scale of a hypothetical outside observer. We consequently discard the above possibility and define *geometric equilibrium* by the requirement

$$\nabla_{(\Gamma} Y_{\Sigma)} = 0. \tag{14.1}$$

Let us summarize our findings thusfar. A world tube $\mathscr{S}$ is the space-time history of an isopleth in the state of geometric equilibrium if and only if there exists a vector field $\{Y^\Gamma\}$ on the three-dimensional, hyperbolic normal metric space $\mathscr{S}^*$ with metric tensor $\{a_{\Gamma\Sigma}\}$ such that

$$Y^\Gamma Y^\Sigma a_{\Gamma\Sigma} = \exp(-2\Psi(u^A)), \qquad Y_{[\Gamma} \nabla_\Sigma Y_{\Lambda]} = 0, \qquad \nabla_{(\Gamma} Y_{\Sigma)} = 0. \tag{14.2}$$

Under satisfaction of these conditions, all observers riding with the isopleth have calibrated and synchronized proper time scales and the space and time measurements on the isopleths do not change in the course of time. It must be clearly noted that the hypothetical observers attached to the isopleths do not necessarily move with any material "particles" that may be thought to comprise the isopleth, for the trajectories of the observers are Fermi-irrotational while the trajectories of the material particles need not have this property.

15. Implications of Geometric Equilibrium

If a hyperbolic-normal metric space with metric tensor $\{a_{\Gamma\Sigma}\}$ satisfies the conditions stated by (14.2), it is known as a static space[15]. Consequently, an $\mathscr{S}^*$ in geometric equilibrium is static. This fact does not imply, however, that the four-dimensional space $\mathscr{E}$ in which $\mathscr{S}^*$ is imbedded as a regular time-like hypersurface is static, as shown by Edelen and Thomas[16].

The properties of static $\mathscr{S}^*$'s are well known by direct analogy with static Einstein-Riemann spaces[17]. Let $\{U^\Gamma(u^A)\}$ be a vector field that is defined on $\mathscr{S}^*$ by the relations

$$U^\Gamma = Y^\Gamma \exp(\Psi(u^A)). \tag{15.1}$$

[14] See Edelen, D. G. B.: *Nuovo Cimento 43 A*, 1095 (1966) for an account of the properties of space-times wherein only the projection of the Lie derivate of the metric tensor with respect to the tangent vector field of a normal congruence must vanish (semistatic space-times).

[15] Lichnerowicz, A.: *Théories Relativistes de la Gravitation.* Paris: Masson 1955.

[16] Edelen, D. G. B., Thomas, T. Y.: *J. Math. Anal. Appl. 7*, 247 (1963).

[17] Lichnerowicz, A.: *Théories Relativistes de la Gravitation.* Paris: Masson 1955; Witten, L. ed.: *Gravitation: An Introduction to Current Research.* New York: John Wiley & Sons 1962.

We then have

$$K_{\Gamma\Sigma} U^{\Sigma} = (a^{\Sigma\Lambda} \nabla_{\Sigma} \nabla_{\Lambda} \Psi) U_{\Gamma}, \tag{15.2}$$

so that $\{U^{\Gamma}\}$ is an eigenvector of the Ricci tensor, $\{K_{\Gamma\Sigma}\}$, of $\mathscr{S}^*$ with the associated eigenvalue $a^{\Sigma\Lambda} \nabla_{\Sigma} \nabla_{\Lambda} \Psi$, and we also have

$$\begin{aligned} &U_{\Gamma} U^{\Gamma} = 1, \quad \nabla_{\Gamma} U_{\Sigma} = U_{\Gamma} \partial_{\Sigma} \Psi, \quad U^{\Gamma} \partial_{\Gamma} \Psi \stackrel{\text{def}}{=} \dot{\Psi} = 0, \\ &\dot{U}_{\Gamma} = U^{\Sigma} \nabla_{\Sigma} U_{\Gamma} = \partial_{\Gamma} \Psi. \end{aligned} \tag{15.3}$$

The proper time trajectories, $\mathscr{K}$, on $\mathscr{S}^*$ were shown to be solutions of the differential equations $\mathrm{d}u^{\Gamma}/\mathrm{d}p = Y^{\Gamma}(u^{\Lambda})$. Since $\{Y^{\Gamma}\}$ is not a unit vector field, it can not serve as the velocity vector field of the observers attached to the isopleth. If, however, we use (15.1) to write the equations for the proper time trajectories in terms of the vector field $\{U^{\Gamma}\}$ and note that

$$\mathrm{d}\Psi/\mathrm{d}p = \partial_{\Gamma} \Psi (\mathrm{d}u^{\Gamma}/\mathrm{d}p) = Y^{\Gamma} \partial_{\Gamma} \Psi = \mathrm{e}^{\Psi} U^{\Gamma} \partial_{\Gamma} \Psi = \mathrm{e}^{\Psi} \dot{\Psi} = 0,$$

by the third equation of the system (15.3), we may write

$$\mathrm{d}(\mathrm{e}^{-\Psi} u^{\Gamma})/\mathrm{d}p = U^{\Gamma}(u^{\Lambda}).$$

Since Ψ is independent of p by the previous equation, we alternatively have

$$\mathrm{d}u^{\Gamma}/\mathrm{d}q = U^{\Gamma}(u^{\Lambda}), \qquad q = \mathrm{e}^{\Psi} p + \text{constant}. \tag{15.4}$$

Thus, as $\{U^{\Gamma}\}$ is a unit timelike vector field on $\mathscr{S}^*$, it may be identified with the velocity vector field of the observers attached to the isopleth and the parameter q may be identified with the proper time of the observers [18].

There is one further property of static $\mathscr{S}^*$'s that we shall need. As is well known [19], the timelike character of the vector field $\{Y^{\Gamma}\}$ together with its continuity and differentiability, insures the existence of local coordinate systems on $\mathscr{S}^*$ such that

$$Y^{\Gamma} \stackrel{*}{=} \delta_0^{\Gamma}. \tag{15.5}$$

Such coordinate systems will be referred to as $*$-coordinate systems, and the symbol $\stackrel{*}{=}$ will be used to denote equalities that obtain in such coordinate systems on $\mathscr{S}^*$. From the definition of the associated covariant components of $\{Y^{\Gamma}\}$, we have $Y_{\Gamma} \stackrel{*}{=} a_{0\Gamma}$. The first of (14.2) thus gives

$$a_{00} \stackrel{*}{=} \exp(-2\Psi). \tag{15.6}$$

Further, when (15.5) is substituted into the remaining equations of the system (14.2), we have

$$\partial_0 a_{\Gamma\Sigma} \stackrel{*}{=} 0, \qquad a_{0\beta} \stackrel{*}{=} 0. \tag{15.7}$$

---

[18] From the first of (15.3) and (15.4) we have $\mathrm{d}q^2 = a_{\Gamma\Sigma} \mathrm{d}u^{\Gamma} \mathrm{d}u^{\Sigma} = \mathrm{d}l^2$ when the differentials are evaluated on the trajectories of $\mathscr{K}$.

[19] Goursat, E.: *Leçons sur le problème de Pfaff*. Paris: 1922.

A $*$-coordinate system on $\mathscr{S}^*$ is thus a very useful construct, for we know that in such coordinate systems the values of $a_{0\Gamma}$ are prescribed and that all of the a's are independent of u^0. Lastly, (15.1) and (15.5) give us

$$U^\Gamma \stackrel{*}{=} \delta_0^\Gamma \exp(\Psi), \qquad U_\Gamma \stackrel{*}{=} \delta_\Gamma^0 \exp(-\Psi) \tag{15.8}$$

for the velocity vector field of the observers attached to the isopleth.

B. Physical Isopleths and Jump Discontinuities

Although we have derived the conditions for the geometric equilibrium of an isoplethic surface in terms of the geometry of its world tube in the 4-dimensional space-time of general relativity, we have not really identified the isoplethic surface with any specific physical construct – the conditions for geometric equilibrium could be applied to any timelike hypersurface whether or not it happens to be the world tube of a physical isoplethic surface. We now have to face the problem of giving a physical identification which singles out isoplethic surfaces from arbitrary timelike hypersurfaces.

One way of obtaining the physical and geometric properties of an isoplethic surface is to specify the momentum-energy tensor that characterizes a galaxy, solve the Einstein field equations, and then examine the surfaces that are the loci of constant values of the salient physical parameters (density, pressure, etc.) that enter into the specification of the momentum-energy tensor. Unfortunately, this is an almost insurmountable task for even the most superficial specification of the momentum-energy tensor. For the momentum-energy tensors that are characteristic of actual galaxies, the labor would be out of the question even if and when we have sufficient knowledge to specify the momentum-energy tensor with any degree of certainty.

The alternative we select to follow is that the isopleth carries a step discontinuity in at least one component of the momentum-energy tensor. In this way, the isoplethic surface is directly identifiable in terms of the physical parameter or parameters that exhibit the step discontinuities, such as in the shell models of elliptical galaxies[20]. Further, as will be shown in this Part, such a discontinuity hypersurface is governed by certain existence conditions for the Einstein field equations, and these conditions in turn imply certain geometric constraints whereby the conditions of geometric equilibrium are simply implemented. In effect, this method allows us to concentrate on the world tube of the isopleth without having to worry about the interior and the exterior solutions of

[20] Schmidt, M.: *Bull. Astr. Inst. Netherlands 13*, 151 (1957).

the Einstein field equations, for the existence of such interior and exterior solutions is assured locally by the method.

Physically, we picture the region interior to $\mathscr{S}$ as being occupied by the world lines of a galactic structure, and the region exterior to $\mathscr{S}$ as containing a background field of what we may think of as "free space" or as an external enveloping environment. For a complete description of this picture we would require a precise statement of the momentum-energy tensor appropriate to the dynamical processes interior and exterior to such a galactic structure. Current knowledge of galaxies is such, however, that we can not specify the momentum-energy tensor for the interior of $\mathscr{S}$. A similar situation obtains in the case of the external or enveloping environment, as was pointed out in Part A of the previous Chapter. The discontinuity that $\mathscr{S}$ is posited to carry in the components of the momentum-energy tensor may accordingly be thought of as providing an approximate description of a transition region between the exterior or cosmological environment, as obtained in Part A of Chapter I, and the internal consistency environment, as obtained in Part B of Chapter I. This situation is analogous to the surface of transition in the classical macroscopic description of galaxies that separates the Boltzmann and the Smolokowski (Knudsen) statistical regimes[21]. In conjunction with the view of perception and cognition given in the Prologue, the introduction of discontinuity hypersurfaces in the modeling of isoplethic surfaces of galaxies gives a concrete realization of galaxies in terms of a threshold of discrimination whereby the galaxy is perceived as a distinct entity. This method also affords a direct realization of the closure property of the threshold perception process and allows the investigator to consider a galaxy as a localized entity rather than a diffuse quantity that extends indefinitely.

Positing jump discontinuities in the momentum-energy tensor carries along with it an equivalent positing of jump discontinuities in the geometric quantities if the Einstein field equations are to have meaning in the presence of such jump discontinuities. We accordingly delve into the requisite surface geometry and discontinuity theory at the beginning of this Part and then return to the main stream of the investigation.

16. Surface Quantities and Continuity Assumptions

If $\mathscr{S}$ carries jump discontinuities in the components of the momentum-energy tensor of $\mathscr{E}$, the Einstein field equations require corresponding jump discontinuities in at least the second derivatives of the components

---

[21] Zwicky, F.: *Morphological Astronomy*. Berlin-Göttingen-Heidelberg: Springer 1957.

of the metric tensor of $\mathscr{E}$ on $\mathscr{S}$. Accordingly, we must spell out just exactly what continuity and differentiability conditions are to be assumed in the theory.

Let P be an arbitrary point in $\mathscr{S}$. We assume the existence of at least one four-dimensional neighborhood $\mathscr{J}(P)$ of P in $\mathscr{E}$ that is divided into two four-dimensional parts by $\mathscr{S}$. Denote by $\mathscr{J}_1(P)$ the subregion of $\mathscr{E}$ that is comprised of all points of $\mathscr{J}(P)$ that lie on one side of $\mathscr{S}$ and by $\mathscr{J}_2(P)$ the subregion of $\mathscr{E}$ that is comprised of all points of $\mathscr{J}(P)$ that lie on the other side of $\mathscr{S}$. Further, let $\mathscr{D}_1(P)$ denote the domain $\mathscr{J}_1(P) + \mathscr{S} \cap \mathscr{J}(P)$, the domain $\mathscr{J}_2(P) + \mathscr{S} \cap \mathscr{J}(P)$ being denoted by $\mathscr{D}_2(P)$. The following assumptions are now made.

A_1: The functions $\{f^A(u^\Gamma)\}$ that appear in the parametric equations of $\mathscr{S}$ are of class C^3 and are such that rank $(x^A_\Gamma) = 3$ at all points of $\mathscr{S}$.
A_2: For every point P of $\mathscr{S}$, the components of the metric tensor of $\mathscr{E}$ are functions of class C^1 in $\mathscr{J}(P)$, of class C^2 in $\mathscr{J}_1(P)$ and in $\mathscr{J}_2(P)$, and have either a right-hand or a left-hand directional derivative of class C^2 in $\mathscr{D}_1(P)$ and in $\mathscr{D}_2(P)$.

It follows from A_2 that the Christoffel symbols of $\mathscr{E}$ are continuous across $\mathscr{S}$, and hence the general relativistic gravitational forces are continuous across $\mathscr{S}$. We thus have a situation similar to that encountered in Newtonian mechanics when a boundary surface of a gravitating body is crossed. Such circumstances are usually designated as those of *second order discontinuity* since the primary mathematical variables (the components of the metric tensor) are assumed to have continuous first derivatives.

Assumption A_1 and (13.1) show that the components of $\{a_{\Gamma\Sigma}\}$ are continuous on $\mathscr{S}$ and have continuous first partial derivatives. The Christoffel symbols formed from $\{a_{\Gamma\Sigma}\}$ are thus continuous functions of the surface coordinates of $\mathscr{S}$. One can thus construct first surface covariant derivatives of differentiable tensorial quantities defined on $\mathscr{S}$[22]. Thus, in particular, we have

$$\nabla_\Gamma x^A_\Sigma = \partial_\Gamma x^A_\Sigma - x^A_\Lambda \Gamma^\Lambda_{\Gamma\Sigma} + x^B_\Sigma x^C_\Gamma \Gamma^A_{BC}.$$

We have already defined the symbols used to designate the coefficient tensors of the first and second fundamental forms on $\mathscr{S}$ and the field of normal vectors to $\mathscr{S}$. A few further details are now needed. Since $\mathscr{S}$ is timelike, its unit normal field $\{N_A\}$ satisfies the equations

$$x^A_\Sigma N_A = 0, \quad N_A N^A = -1. \tag{16.1}$$

---

[22] Conceptually, it is much easier to consider all differentiations on the surface $\mathscr{S}$ as ordinary differentiations in the metric space $\mathscr{S}^*$.

The first fundamental form gives rise to the relations[23]

$$a_{\Gamma\Sigma} = h_{AB}(f^C)x^A_\Gamma x^B_\Sigma, \quad a^{\Gamma\Sigma}x^A_\Gamma x^B_\Sigma = h^{AB}(f^C) + N^A N^B. \tag{16.2}$$

The coefficient tensor of the second fundamental form is defined by

$$b_{\Gamma\Sigma} = -(\nabla_\Gamma x^A_\Sigma)N_A, \tag{16.3}$$

and gives rise to the relations

$$\nabla_\Gamma x^A_\Sigma = b_{\Gamma\Sigma}N^A, \quad \nabla_\Gamma N^A = b_{\Gamma\Sigma}a^{\Sigma\Lambda}x^A_\Lambda. \tag{16.4}$$

17. Jump Strengths and Existence Conditions

So far the hypersurface $\mathscr{S}$ has been any regular, timelike hypersurface in $\mathscr{E}$. We now specifically restrict our attention to those hypersurfaces that carry basic field-theoretic information in the sense that they are the support hypersurfaces for field discontinuities. In view of the assumptions A_1 and A_2, this is accomplished by the following requirement.

A_3: There is a jump discontinuity in at least one of the second derivatives of the functions $h_{AB}(x^K)$ at points of $\mathscr{E}$ that lie on $\mathscr{S}$.

The symbol $[\![W^{A\cdots}_{B\cdots}]\!]$ will be used to denote the jumps in the quantities $W^{A\cdots}_{B\cdots}$ across $\mathscr{S}$ (the value of a quantity as $\mathscr{S}$ is approached from the "outside" diminished by the value of a quantity as $\mathscr{S}$ is approached from the "inside"). Assumption A_2 gives us

$$[\![h_{AB}]\!] = 0, \quad [\![\partial_C h_{AB}]\!] = 0, \tag{17.1}$$

while A_3 states that $[\![\partial_C\partial_D h_{AB}]\!] \neq 0$ for some choice of the indices. In fact, it can be shown[24] that there exist functions $\lambda_{AB}(u^\Gamma)$ defined on $\mathscr{S}$ with the property

$$[\![\partial_C\partial_D h_{AB}]\!] = \lambda_{AB}N_C N_D. \tag{17.2}$$

---

[23] The basic results from the differential geometry of hypersurfaces is used without further notice. The reader is referred to Schouten: p. 242, or to Thomas, T. Y.: *J. Math. Anal. Appl.* 7, 225 (1963) for the particular aspects as they apply here. Thomas' results can be used directly, Schouten's results must be modified to account for the fact that the first fundamental form is not positive definite.

[24] The results of this Section are reasonably well reported in the literature, in contrast to those of the next Section: Stellmacher, K.: *Math. Ann. 115*, 750 (1938); O'Brien, S., Synge, J. L.: *Comm. Dublin Inst. Adv. Studies Ser. A9* (1953); Papapetrou, A., Treder, H.: *Math. Nachr. 20,* 53 (1959); Dautcourt, G.: *Arch. Rational Mech. Anal. 13*, 55 (1963); Treder, H.: *Gravitative Stoßwellen*. Berlin: 1962; Edelen, D. G. B., Thomas, T. Y.: *Arch. Rational Mech. Anal. 9,* 153, 245 (1962); Edelen, D. G. B., Thomas, T. Y.: *J. Math. Anal. Appl. 7,* 247 (1963); Thomas, T. Y.: *J. Math. Anal. Appl. 7,* 225 (1963).

In view of (17.1) and the fact that

$$\lambda_{AB} = [\![\partial_C \partial_D h_{AB}]\!] N^C N^D, \tag{17.3}$$

the λ's are referred to as the *jump strengths* of the metric field.

We now impose the requirements that the structure of the space $\mathscr{E}$ is determined by the Einstein field equations

$$R_{AB}(h) - \tfrac{1}{2} R(h) h_{AB} = \kappa T_{AB}. \tag{17.4}$$

If these equations are to hold in the presence of second order discontinuities, certain conditions must be fulfilled. These conditions are in many ways similar to the Rankine-Hugoniot relations of fluid mechanics[25]. Let us denote the *jump strengths* of the momentum-energy tensor across $\mathscr{S}$ by $\{S_{AB}\}$; that is,

$$S_{AB} = [\![T_{AB}]\!] \tag{17.5}$$

where the S's are functions of the surface coordinates on $\mathscr{S}$. It can then be shown that necessary conditions for the *existence* of solutions of the Einstein field equations under assumptions A_1, A_2 and A_3 are that the discontinuity strengths satisfy the equations

$$[\![R_{AB}(h)]\!] - \tfrac{1}{2}[\![R(h)]\!] h_{AB} = \kappa S_{AB}. \tag{17.6}$$

Since both $\{R_{AB}(h)\}$ and $R(h)$ are second order differential operators on $\{h_{AB}\}$ that are linear in the second derivatives, a straightforward calculation based on (17.1), (17.2) and (17.6) leads to the requirements

$$\lambda_{AB} + 2 Z_{(A} N_{B)} - \lambda_{CD} h^{CD} N_A N_B - (Z + \lambda_{CD} h^{CD}) h_{AB} = 2\kappa S_{AB}, \tag{17.7}$$

where

$$Z_A = \lambda_{AB} N^B, \qquad Z = Z_A N^A. \tag{17.8}$$

If we multiply (17.7) by $\{N^B\}$, sum on the repeated index, and use (17.8), the left members of the resulting equations vanish identically. We are thus left with the simple set of relations

$$S_{AB} N^B = 0 \tag{17.9}$$

which are the well-known Synge-O'Brien junction conditions. It is also evident that the functions $\{Z_A\}$ are undetermined by the Eqs. (17.7), (17.8) and hence may be taken as arbitrary functions of the coordinates of $\mathscr{S}$. In addition, if there are no physical jumps (i. e., $S_{AB} = 0$), the λ's are solely dependent upon the Z's and the admissible choice $Z_A = 0$ anihilates the λ's. In this sense, the metrical jump strengths determined by the Z's have no intrinsic physical meaning although they do become objects of study in the analysis of free gravitational radiation problems.

---

[25] Taub, A. H.: *Phys. Rev. 74*, 328 (1948); – *Illinois J. Math. 1*, 370 (1957); Saini, G. L.: *J. Math. Mech. 10*, 887 (1961).

The fundamental system of existence conditions (17.7) and (17.9) involve the jump strengths $\{\lambda_{AB}\}$ and $\{S_{AB}\}$, which are tensor quantities under admissible coordinate transformations in the four-dimensional space $\mathscr{E}$. Now, consider the quantities $\{S_{\Gamma\Sigma}\}$ defined by the relations

$$S_{\Gamma\Sigma} = S_{AB} x^A_\Gamma x^B_\Sigma . \tag{17.10}$$

Under admissible transformations of the surface coordinates on $\mathscr{S}$, these quantities transform according to the indicated tensor law, and hence constitute the components of a surface tensor field. A direct calculation based on (16.2) and (17.9) gives

$$S^{AB} = S^{\Gamma\Sigma} x^A_\Gamma x^B_\Sigma , \tag{17.11}$$

where $\{S^{\Gamma\Sigma}\}$ is obtained from $\{S_{\Gamma\Sigma}\}$ when the indices are raised by means of $\{a^{\Gamma\Sigma}\}$ in the usual manner. On the other hand, if we assume (17.11), Eqs. (17.9) are identically satisfied. We thus see that the surface quantities $\{S_{\Gamma\Sigma}\}$ give a unique determination of $\{S_{AB}\}$ and contain all relevent algebraic information, for one may also verify that

$$S = S_{AB} h^{AB} = S_{AB}(a^{\Gamma\Sigma} x^A_\Gamma x^B_\Sigma - N^A N^B) = S_{\Gamma\Sigma} a^{\Gamma\Sigma} . \tag{17.12}$$

We now consider the surface tensor whose components are defined by

$$\lambda_{\Gamma\Sigma} = \lambda_{AB} x^A_\Gamma x^B_\Sigma . \tag{17.13}$$

If (17.7) is written in the equivalent form

$$\lambda_{AB} = 2\kappa S_{AB} - 2Z_{(A} N_{B)} - (Z + \kappa S) N_A N_B - \kappa S h_{AB} ,$$

and we transvect both sides with $\{x^A_\Gamma x^B_\Sigma\}$, we obtain the following simple system of surface equations:

$$\lambda_{\Gamma\Sigma} = \kappa(2S_{\Gamma\Sigma} - S a_{\Gamma\Sigma}) . \tag{17.14}$$

It is easily seen that the four-dimensional specification of the metric jump strengths can be written in the form

$$\lambda^{AB} = \lambda^{\Gamma\Sigma} x^A_\Gamma x^B_\Sigma - 2Z^{(A} N^{B)} - Z N^A N^B \tag{17.15}$$

by noting that $\{x^A_\Gamma\}$ and $\{N^A\}$ form a basis for the contravariant vector space of $\mathscr{E}$ at any point on $\mathscr{S}$ and making use of (17.8). A simple calculation then shows that when $\{\lambda_{\Gamma\Sigma}\}$ is derived from (17.14), the system (17.15) implies the identical satisfaction of the existence conditions (17.7). Further, we have

$$\lambda^{AB} h_{AB} = \lambda - Z , \qquad \lambda = a^{\Gamma\Sigma} \lambda_{\Gamma\Sigma} . \tag{17.16}$$

A combination of the above considerations leads us to the following basic result. *Necessary conditions for the existence of solutions of the Ein-*

stein field equations under assumptions A_1, A_2 and A_3 are given by the surface tensorial equations

$$\lambda_{\Gamma\Sigma} = \kappa(2S_{\Gamma\Sigma} - S a_{\Gamma\Sigma}). \tag{17.17}$$

The quantities $\{S_{\Gamma\Sigma}\}$ and $\{\lambda_{\Gamma\Sigma}\}$ then serve to determine the jump strengths $\{S_{AB}\}$ and $\{\lambda_{AB}\}$ by (17.10) *and* (17.15). By these results we are permitted to base our succeeding considerations on surface tensors and the intrinsic geometry of $\mathscr{S}$. This is indeed an economy, for it reduces the problem from a four-dimensional one to an equivalent three-dimensional one, at least as far as the necessary relations that the jump strengths must satisfy.

18. Differential Relations

The previous Section has shown that knowledge of the surface quantities $\{S_{\Gamma\Sigma}\}$ is sufficient to determine $\{\lambda_{\Gamma\Sigma}\}$ and hence $\{\lambda_{AB}\}$ to within the geometrical structure of $\mathscr{S}$, i.e., to within the quantities $\{a_{\Gamma\Sigma}\}$, $\{N_A\}$, $\{x^A_\Gamma\}$, and $\{h_{AB}\}$. As yet, however, we have not used the full content of the Einstein theory, for the relations

$$\nabla_B T^B_A = 0 \tag{18.1}$$

are still at our disposal. We shall now show[26] that the system (18.1) leads to a set of differential relations on $\mathscr{S}$ (in $\mathscr{S}^*$) that will partially determine the structure of $\{S_{\Gamma\Sigma}\}$ and the geometry of $\mathscr{S}$.

We first note that, by (17.5), we have

$$\nabla_\Gamma S^B_A = \nabla_\Gamma [\![T^B_A]\!] = [\![\nabla_C T^B_A]\!] x^C_\Gamma . \tag{18.2}$$

If we transvect (18.2) with $\{x^D_\Sigma a^{\Sigma\Gamma}\}$ and use (16.2), we are led to the equations

$$(\nabla_\Gamma S^B_A) a^{\Gamma\Sigma} x^D_\Sigma = [\![\nabla_C T^B_A]\!] (h^{CD} + N^C N^D). \tag{18.3}$$

Obvious manipulations of (18.3) then give

$$[\![\nabla_C T^B_A]\!] = (\nabla_\Gamma S^B_A) a^{\Gamma\Sigma} x^D_\Sigma h_{CD} - F^B_A N_C, \tag{18.4}$$

where

$$F^B_A = [\![\nabla_C T^B_A]\!] N^C \tag{18.5}$$

are functions of the coordinates of the hypersurface $\mathscr{S}$. Now, since (18.1) holds on both sides of $\mathscr{S}$, the tensor equations

$$[\![\nabla_B T^B_A]\!] = 0 \tag{18.6}$$

must hold on $\mathscr{S}$. Substituting (18.4) into (18.6), we obtain the differential relations

$$(\nabla_\Gamma S_{AB}) x^B_\Sigma a^{\Gamma\Sigma} = F_A, \tag{18.7}$$

---

[26] The analysis follows Edelen, D. G. B., Thomas, T. Y.: *J. Math. Anal. Appl. 7,* 247 (1963).

where the quantities $\{F_A\}$ constitute the components of a vector in $\mathscr{E}$ restricted to $\mathscr{S}$ that is given by

$$F_A = [\![\nabla_C T_{AB}]\!] N^B N^C = [\![\nabla_C T_{AB}]\!] x^B_\Gamma x^C_\Sigma a^{\Gamma\Sigma}. \tag{18.8}$$

We now replace the differential relations (18.7) by an equivalent set that involves $\{S_{\Gamma\Sigma}\}$ rather than $\{S_{AB}\}$, in accordance with the results established at the end of the previous Section. For this purpose, we transvect (18.7) with $\{N^A\}$ and obtain

$$N^A(\nabla_\Gamma S_{AB}) x^B_\Sigma a^{\Gamma\Sigma} = N^A F_A \stackrel{\text{def}}{=} \chi. \tag{18.9}$$

By covariant surface differentiation of (17.9) and use of (16.4), we find that (18.9) is equivalent to

$$S^{\Gamma\Sigma} b_{\Gamma\Sigma} + \chi = 0. \tag{18.10}$$

On the other hand, if we transvect (18.7) with $\{x^A_\Lambda\}$, we have

$$(\nabla_\Gamma S_{AB}) x^B_\Sigma x^A_\Lambda a^{\Gamma\Sigma} = F_A x^A_\Lambda \stackrel{\text{def}}{=} F_\Lambda. \tag{18.11}$$

It then follows from our previous relations that

$$\begin{aligned} \nabla_\Gamma S^\Gamma_\Lambda &= \nabla_\Gamma (S_{AB} x^A_\Lambda x^B_\Sigma a^{\Gamma\Sigma}) \\ &= F_\Lambda + S_{AB}(b_{\Gamma\Lambda} N^A x^B_\Sigma + b_{\Gamma\Sigma} N^B x^A_\Lambda) a^{\Gamma\Sigma} = F_\Lambda. \end{aligned}$$

The following fundamental result thus obtains since $\{x^A_\Gamma\}$ and $\{N^A\}$ constitute a basis for the contravariant vector space of $\mathscr{E}$ restricted to $\mathscr{S}$. *Necessary conditions for the existence of solutions to the Einstein field equations under assumptions A_1, A_2 and A_3 are given by the relations*

$$S^{\Gamma\Sigma} b_{\Gamma\Sigma} + \chi = 0, \tag{18.12}$$

$$\nabla_\Gamma S^\Gamma_\Sigma = F_\Sigma. \tag{18.13}$$

The above relations being consequences of the relativistic equations of motion, (18.1), in the presence of jump discontinuities in the momentum-energy tensor, χ may be interpreted as the jump in the normal component of the force across $\mathscr{S}$, while $\{F_\Sigma\}$ assumes the role of a shear force on $\mathscr{S}$. These interpretations are also immediate consequences of (18.8), (18.9) and (18.11) on noting that $\{F_A\}$ is the jump strength of the force field across $\mathscr{S}$ in $\mathscr{E}$ that is required in order to equilibrate the momentum-energy jump strengths.

Although it is of no direct interest in this theory, we note that the above results may be used to obtain differential and algebraic equations for the surface quantities $\{\lambda_{\Gamma\Sigma}\}$:

$$\lambda^{\Gamma\Sigma} b_{\Gamma\Sigma} - \lambda a^{\Gamma\Sigma} b_{\Gamma\Sigma} + 2\kappa\chi = 0, \tag{18.15}$$

$$\nabla_\Gamma \lambda^\Gamma_\Sigma - \partial_\Sigma \lambda = 2\kappa F_\Sigma. \tag{18.16}$$

19. The Continuation Problem

If the components of the momentum-energy tensor are completely known on both sides of $\mathcal{S}$, the quantities $\{S_{\Gamma\Sigma}\}$ can be determined by (17.5) and (17.11). Accordingly, if the Einstein field equations are to be solvable in any neighborhood that intersects $\mathcal{S}$, the quantities χ and $\{F_A\}$ must be such that (18.12) and (18.13) reduce to algebraic identities at every point of $\mathcal{S}$. On the other hand, if the momentum-energy tensor is not completely known throughout even one neighborhood that intersects $\mathcal{S}$ and if the Einstein field equations are posited to be solvable, (18.13) becomes a differential system that must be solved for $\{S_{\Gamma\Sigma}\}$ subject to the algebraic constraint stated by (18.12). This situation constitutes the so-called continuation problem whereby general solutions of the Einstein field equations may be continued across a surface of jump discontinuity in the momentum-energy tensor when there is incomplete knowledge of the momentum-energy tensor on at least one side of $\mathcal{S}$[27]. Admittedly, we have, at present, incomplete knowledge concerning the momentum-energy tensor of a galaxy, even to the point where it is not evident that the "perfect fluid" model is adequate as a first approximation. It thus seems preferable to let the "chips fall where they may" and not to assume that the $\{S_{\Gamma\Sigma}\}$ are known *a priori*. We adhere to this viewpoint throughout, even though an incomplete specification of the momentum-energy tensor leads to difficulties in establishing simple physical interpretations. Be this as it may, an honest admission of ignorance of the physical processes of galaxies is preferable to a facile analysis that bristles with elemental physical assumptions.

A complete solution of the continuation problem provides all admissible momentum-energy jump strengths for which the Einstein field equations are solvable. This follows from the fact that (18.12) and (18.13) are *existence conditions* on the $\{S_{\Gamma\Sigma}\}$ and hence on the $\{S_{AB}\}$, and these quantities in turn determine the physical junction conditions

$$T_{AB}|_{\mathcal{S}\cap\mathcal{D}_1} = T_{AB}|_{\mathcal{S}\cap\mathcal{D}_2} + S_{AB}$$

that must be used in solving the equations of motion $\nabla_B T^B_A = 0$. The corresponding geometric junction conditions are given by

$$\partial_C \partial_D h_{AB}|_{\mathcal{S}\cap\mathcal{D}_1} = \partial_C \partial_D h_{AB}|_{\mathcal{S}\cap\mathcal{D}_2} + \lambda_{AB} N_C N_D,$$

where the λ's are determined in terms of the S's and the geometry of $\mathcal{S}$ by (17.15) and (17.17).

---

[27] For an explicit example of such problems, see Edelen, D. G. B.: *J. Math. Anal. Appl. 4,* 346 (1962).

20. Solution of the Continuation Problem

The only information available to us in a continuation problem is that provided by the existence conditions

$$\nabla_\Gamma S^\Gamma_\Sigma = F_\Sigma, \tag{20.1}$$

and

$$S^{\Gamma\Sigma} b_{\Gamma\Sigma} + \chi = 0, \qquad S_{[\Gamma\Sigma]} = 0. \tag{20.2}$$

One possibility would be to specify the S's in an arbitrary fashion and then to determine the $\{F^A\}$, since our previous results easily lead to the determination

$$F^A = S^{\Gamma\Sigma} b_{\Gamma\Sigma} N^A + (\nabla_\Gamma S^{\Gamma\Sigma}) x^A_\Sigma. \tag{20.3}$$

This procedure would amount, however, to a fitting of the equations of motion on one side of $\mathscr{S}$ to those on the other side by an arbitrary specification of the quantities $[\![\nabla_C T^B_A]\!] N^C N_B$, the latter quantities being uniquely determined by the arbitrarily chosen S's. Having observed that such an artificial fitting process is possible, we discard it as a general method of procedure. We are thus left with the problem of determining the S's such that Eqs. (20.1) and (20.2) are satisfied when the functions $\{F_\Sigma\}$ and χ are arbitrary, unspecified functions on the hypersurface $\mathscr{S}$. In this way, we will then be able to accomodate normal and shear forces that may be present on $\mathscr{S}$ without having to have the specific functional dependence of these forces on the coordinates of $\mathscr{S}$.

We first fix our attention on the system (20.1). These equations may be considered as an invariant differential system of the three-dimensional hyperbolic-normal metric space $\mathscr{S}^*$ with coordinates $\{u^\Gamma\}$ and metric differential form

$$dl^2 = a_{\Gamma\Sigma}(u^A) du^\Gamma du^\Sigma. \tag{20.4}$$

If the system (20.1) is to possess a solution in $\mathscr{S}^*$, there must exist at least one set of point-valued functions $\{Q^{\Gamma\Sigma}(u^A)\}$ on $\mathscr{S}^*$ such that they are the components of a symmetric tensor field and that they satisfy the relations

$$\nabla_\Gamma Q^{\Gamma\Sigma} = F^\Sigma \qquad (F^\Sigma = F_A a^{A\Sigma}) \tag{20.5}$$

identically throughout $\mathscr{S}^*$. In the context of differential equations, the tensor $\{Q^{\Gamma\Sigma}\}$ plays the role of a particular solution of the linear differential system (20.1). The linearity of (20.1) allows us to write its general solution in the form

$$S_{\Gamma\Sigma} = Q_{\Gamma\Sigma} + Z_{\Gamma\Sigma}, \tag{20.6}$$

where $\{Z_{\Gamma\Sigma}\}$ constitute a general solution of the associated homogeneous system

$$\nabla_\Gamma Z^{\Gamma\Sigma} = 0, \qquad Z^{[\Gamma\Sigma]} = 0. \tag{20.7}$$

To some extent it would appear that we have "robbed Peter to pay Paul", for we now have to face the problem of determining the functions $\{Z^{\Gamma\Sigma}\}$. Now, the system (20.7) is formally similar to the equations of motion in general relativity since $\mathscr{S}^*$ is a hyperbolic-normal metric space. In addition, the T's represent momentum-energy complexes in $\mathscr{E}$ while the Z's represent part of the jump strengths in the momentum-energy and hence have energetic interpretations in $\mathscr{S}^*$. The formal analogy goes even deeper, for the Z's also partially determine the λ's by (20.7) and (17.17) and the λ's represent jumps in the second coordinate derivatives of the geometric quantities $\{h_{AB}\}$. In this vein, it is also known that the intrinsic geometry of a discontinuity hypersurface enters into the determination of the classical jump strengths of physical quantities in an essential fashion. There is thus a natural basis for expecting that the intrinsic geometry of $\mathscr{S}^*$ will partially determine the S's. Now, the quantities $\{Q^{\Gamma\Sigma}\}$ are a specific collection of point valued functions that constitute a particular solution of (20.1) and hence they depend on the specific physical quantities $\{F_\Sigma\}$. The tensor $\{Q^{\Gamma\Sigma}\}$ is accordingly referred to as the *shear potential* since $\{F_\Sigma\}$ is the shear force on $\mathscr{S}$. Eqs. (20.6) thus show that the explicit dependence of $\{S_{\Gamma\Sigma}\}$ on the intrinsic geometry of $\mathscr{S}^*$ must arise through the quantities $\{Z^{\Gamma\Sigma}\}$. We accordingly state the following postulates for the integration of the system (20.7).

P_1: The quantities $\{Z_{\Gamma\Sigma}\}$ are the components of a metric tensor differential invariant[28] of the space $\mathscr{S}^*$.

P_2: This metric tensor differential invariant is of the second order and is linear in the second derivatives of the components of the metric tensor of $\mathscr{S}^*$.

Postulate P_2 expresses the usual restriction to second-order differential relations that is commonly assumed in physical theories. On the other hand, postulate P_1 is a fundamental statement concerning the relation between the physics and the geometry. By this postulate, we are able to relate the physical information carried by a discontinuity hypersurface and the intrinsic geometry of that hypersurface.

The procedure we must follow is now straightforward. It follows from (20.7) that the divergence of the metric tensor differential invariant $\{Z_{\Gamma\Sigma}\}$ of $\mathscr{S}^*$ must vanish identically. The known procedures for constructing such tensor differential invariants[29] show that the most general quantity which satisfies our requirements is given by

[28] See Thomas, T. Y.: *Differential Invariants of Generalized Spaces*. Cambridge: 1934, for a full discussion of differential invariants.

[29] Thomas, T. Y.: *Differential Invariants of Generalized Spaces*. Cambridge: Cambridge University Press 1934.

$$Z_{\Gamma\Sigma} = \Theta\{R_{\Gamma\Sigma}(a) - \tfrac{1}{2}(R(a) + \Pi)a_{\Gamma\Sigma}\}, \tag{20.8}$$

where Θ and Π may be viewed as integration constants and $\{R_{\Gamma\Sigma}(a)\}$ is the Ricci tensor of the space $\mathscr{S}^*$. The constants of integration are assumed to have arbitrary but fixed values that satisfy the constraint $\Theta \neq 0$. Since $\mathscr{S}^*$ is three-dimensional, $\{R_{\Gamma\Sigma}(a)\}$ and $\{a_{\Gamma\Sigma}\}$ serve to determine uniquely the full curvature tensor of $\mathscr{S}^*$.

We have now established a consistent procedure whereby the S's may be obtained such that we simultaneously satisfy the existence requirement (20.1) and the requirements of postulates P_1 and P_2 when $\{F_\Sigma\}$ is an arbitrarily assigned surface vector field. This leaves the existence condition (20.2) to be implemented. Now, the Q's are a particular point-valued solution of the linear system (20.1), and as such, we may write

$$Q_{\Gamma\Sigma} = W_{\Gamma\Sigma}(F_A) + B_{\Gamma\Sigma} \tag{20.9}$$

where the W's are unique functionals of the F's such that

$$\xi\, W_{\Gamma\Sigma}(F_A) = W_{\Gamma\Sigma}(\xi F_A), \quad \partial_\Gamma \xi = 0 \tag{20.10}$$

is an identity in ξ and the B's are any specific point-valued functions on $\mathscr{S}^*$ for which

$$\nabla_\Gamma B^{\Gamma\Sigma} = 0, \quad B^{[\Gamma\Sigma]} = 0 \tag{20.11}$$

hold. If we substitute (20.6) and (20.9) into (20.2), we obtain

$$B^{\Gamma\Sigma} b_{\Gamma\Sigma} = -\chi - (W^{\Gamma\Sigma} + Z^{\Gamma\Sigma}) b_{\Gamma\Sigma}. \tag{20.12}$$

Since all of the terms on the right-hand side of this equation are known, (20.12) is seen to be an equation for the determination of $\{B_{\Gamma\Sigma}\}$. Combining this with (20.11), we have four equations for the determination of the six independent quantities that comprise $\{B_{\Gamma\Sigma}\}$. Hence, since it is easily demonstrated that (20.11) can be satisfied by other than the zero-tensor, the existence requirements (20.2) can always be satisfied; in fact, we are free to add two more conditions in most cases.

The following conclusions thus obtain. Let the functions χ and $\{F_\Sigma\}$ be arbitrarily assigned functions of the coordinates $\{u^\Gamma\}$ of $\mathscr{S}^*$. Then a consistent procedure for the continuation of $\{T_{AB}\}$ across $\mathscr{S}$, such that (1) postulates P_1 and P_2 are satisfied and (2) the Einstein field equations are solvable under assumptions A_1, A_2 and A_3, is given by

$$S_{\Gamma\Sigma} = Q_{\Gamma\Sigma} + \Theta\{R_{\Gamma\Sigma}(a) - \tfrac{1}{2}(R(a) + \Pi)a_{\Gamma\Sigma}\}, \tag{20.13}$$

$$\Theta R_{\Gamma\Sigma}(a) b_{\Gamma\Sigma} - \tfrac{1}{2}(R(a) + \Pi)a^{\Gamma\Sigma} b_{\Gamma\Sigma}\Theta + Q^{\Gamma\Sigma} b_{\Gamma\Sigma} + \chi = 0, \tag{20.14}$$

$$\lambda_{\Gamma\Sigma} = 2\kappa(\Theta R_{\Gamma\Sigma}(a) + Q_{\Gamma\Sigma}) - \tfrac{1}{2}\kappa\{\Theta(R(a) - \Pi) + 2Q\}a_{\Gamma\Sigma}, \tag{20.15}$$

$$\nabla_\Gamma Q^{\Gamma\Sigma} = F^\Sigma, \tag{20.16}$$

where Θ and Π are constants such that $\Theta \neq 0$.

21. Physical Identification of the Isopleths

Although we have referred to $\mathscr{S}$ as an isoplethic surface, we have really only given $\mathscr{S}$ a physical identification by the requirement that it carries a jump discontinuity in the momentum-energy tensor of $\mathscr{E}$. In order that $\mathscr{S}$ shall be well defined in the physical domain, it is necessary that the nature of the momentum-energy jumps be specified by an identification of certain of the quantities that comprise $\{S_{\Gamma\Sigma}\}$ with specific physical quantities. The hypersurface $\mathscr{S}$ then becomes an isopleth when the physical quantities in $\mathscr{E}$ that experience jumps are such that the jumps occur at specified values of the physical quantities. The purpose of this Section is to implement these additional requirements.

It is widely accepted that a first-order approximation of the dynamical processes of elliptical galaxies is provided by the perfect fluid model. The results established in Chapter I show that both the internal environment and the external environment of galaxies are consistent with a perfect fluid first approximation, for the momentum-energy tensor obtained in both cases has the perfect fluid model as the leading term if any reasonable approximation is employed; this obtains in the case of the internal environment when the commutator tensor is of second order in (12.2), and it obtains in the case of the external environment when the tensor $\{Q^{AB}\}$ is of second order. We will obviously require the additional structure involved in the momentum-energy tensor that was developed in Chapter I in our succeeding considerations. For the time being, however, it is sufficient to consider the perfect fluid approximation in the search for physical identification of the isopleths.

If a galaxy be envisioned as an effectively isolated world cylinder of perfect fluid, it is easily shown[30] that the admissible momentum-energy jumps, $\{S^*_{\Gamma\Sigma}\}$, must have the form

$$S^*_{\Gamma\Sigma} = -\bar{\rho}\, W_\Gamma W_\Sigma, \qquad W_\Gamma = \bar{W}_A x^A_\Gamma, \qquad \bar{W}_A N^A = 0, \qquad \bar{\rho} > 0, \quad (21.1)$$

where $\{\bar{W}^A\}$ is the velocity vector field and $\bar{\rho}$ is the density of the fluid evaluated on the immediate interior of $\mathscr{S}$. In general, the fluid velocity vector $\{W_\Gamma\}$ and the velocity vector $\{U_\Gamma\}$ of an observer attached to the isopleth, whereby geometric equilibrium is established, will be different since $\{U_\Gamma\}$ is an irrotational vector field while $\{W_\Gamma\}$ will generally contain a rotational component. Thus, if we write

$$W_\Gamma = a\, U_\Gamma + v_\Gamma, \qquad a = W_\Gamma U^\Gamma, \qquad v_\Gamma U^\Gamma = 0, \qquad (21.2)$$

and note that both $\{W_\Gamma\}$ and $\{U_\Gamma\}$ are unit timelike vectors in $\mathscr{S}^*$, that is,

$$a^2 = 1 + v^2, \qquad v^2 = -v_\Gamma v^\Gamma \geq 0, \qquad (21.3)$$

---

[30] Edelen, D. G. B.: *Proc. Nat. Acad. Sci. U.S.A. 50*, 469 (1963).

we may write $\{S^*_{\Gamma\Sigma}\}$ in the equivalent form

$$S^*_{\Gamma\Sigma} = -\bar{\rho}(1+v^2)U_\Gamma U_\Sigma - 2\bar{\rho}(1+v^2)^{\frac{1}{2}} U_{(\Gamma} v_{\Sigma)} - \bar{\rho} v_\Gamma v_\Sigma . \tag{21.4}$$

We now identify the isoplethic surface under consideration as the surface on which the mass density of a galaxy achieves the value $\rho_1 = \rho_0 + \bar{\rho}$ above the background mass density ρ_0, and where the dynamics of the galaxy are described to the first order approximation by the perfect fluid model with a *constant* jump discontinuity, $-\bar{\rho}$, in the mass density across the isopleth. We may therefore write

$$S_{\Gamma\Sigma} = S^*_{\Gamma\Sigma} + H_{\Gamma\Sigma} \tag{21.5}$$

where $\{H_{\Gamma\Sigma}\}$ stands for second and higher order terms. In most instances, the modulus of the vector field $\{v_\Gamma\}$ will be very small compared to unity, and hence (21.5) would reduce to

$$S_{\Gamma\Sigma} = -\bar{\rho}\, U_\Gamma U_\Sigma + H^*_{\Gamma\Sigma} \tag{21.6}$$

where $\{H^*_{\Gamma\Sigma}\}$ denotes the composition of all second and higher order terms that result within the order of approximation considered here.

The identification obtained above is an essential one in the theory, and hence it is necessary to examine it in greater detail. This is most conveniently done by use of the results obtained in Part A of Chapter I. Since $\{g_{AB}\}$ is continuous with continuous first and second partial derivatives, (17.1)–(17.3) and (2.1) yield

$$[\![\psi]\!] = [\![\partial_A \psi]\!] = 0\,, \quad [\![\partial_C \partial_D \psi]\!] = J N_C N_D\,, \quad \lambda_{AB} = J h_{AB}\,, \quad J = [\![\partial_C \partial_D \psi]\!] N^C N^D\,, \tag{21.7}$$

and accordingly, we have

$$\lambda_{\Gamma\Sigma} = J a_{\Gamma\Sigma}\,. \tag{21.8}$$

It thus follows from (17.5) that a jump discontinuity in the second partial derivatives of the conformal coefficient of a conformally homogeneous model requires that

$$\bar{S}_{\Gamma\Sigma} = -\kappa^{-1} J a_{\Gamma\Sigma}\,, \tag{21.9}$$

where the $\bar{S}$'s denote the jump strengths that obtain in such models. The corresponding jumps in the physical quantities that comprise $\{'T_{AB}\}$ can then be calculated from the results given in Chapter I. These calculations are quite protracted, however, since the jump in a product is generally not equal to the product of the jumps. In order to eliminate this complication, we consider the situation in which the conformal factor vanishes on the exterior of $\mathscr{S}$ in $\mathscr{E}$. It is then an easy matter to show that

$$\kappa [\![' \rho]\!] = -J\,. \tag{21.10}$$

When this result is compared with (21.9), we obtain a form similar to that given by (21.6) with the exception that the terms denoted by $\{H^*_{\Gamma\Sigma}\}$ in (21.6) are now of the same order as the other terms. A little reflection on the forms of the results given in Chapter I and (21.7) shows that (21.10) provides a reasonably accurate description provided we are sufficiently far from the nucleus of a galaxy that the anisotropies and inhomogeneities are relatively insignificant. The identification given in the first part of this Section is thus consistent with the external enveloping environment. Further, the external environment requires satisfaction of (21.9) in any event, and this result is not contingent on the assumption that the conformal coefficient vanish on the exterior of $\mathscr{S}$. This will be of importance in what follows.

C. Physical Isopleths in Geometric Equilibrium

Conditions for the geometric equilibrium of an isoplethic surface were obtained in Part A in terms of the intrinsic geometry of the world tube of the isopleth. We obtained a physical characterization of the isopleth in Part B as a surface that supports a jump discontinuity in the momentum-energy tensor of a form such that it is representative of a jump in the mass density of a galaxy above a fixed background field. Since this characterization was independent of any considerations of equilibrium, the dynamics of the world tube of the physical isopleth is unconstrained at this point to within the existence conditions (18.12) and (18.13). This Part combines the characterization of physical isopleths and the considerations of geometric equilibrium, and yields the general theory of equilibrium galactic structures to which this Tract is addressed. From this we will be able to obtain properties of physical isopleths that are in a state of geometric equilibrium.

22. Determination of $\{S_{\Gamma\Sigma}\}$

Our starting point is the basic solution of the continuation problem given by (20.13). If we solve this system of equations for $\{R_{\Gamma\Sigma}(a)\}$, we obtain

$$\Theta R_{\Gamma\Sigma}(a) = S_{\Gamma\Sigma} - Q_{\Gamma\Sigma} - (S - Q + \Theta \Pi) a_{\Gamma\Sigma}, \tag{22.1}$$

where, as usual, $S = S_{\Gamma\Sigma} a^{\Gamma\Sigma}$ and $Q = Q_{\Gamma\Sigma} a^{\Gamma\Sigma}$. If the world tube of the isopleth is in a state of geometric equilibrium, $\mathscr{S}^*$ must be a static three-dimensional space and hence $\{R_{\Gamma\Sigma}(a)\}$ must satisfy the conditions stated by (15.2). Accordingly, by (15.2) and (22.1), a physical isopleth in a state of geometric equilibrium must satisfy the requirements

$$(S_{\Gamma\Sigma} - Q_{\Gamma\Sigma}) U^{\Sigma} = \mu U_{\Gamma}, \tag{22.2}$$

$$\mu = \Theta a^{\Gamma\Sigma} \nabla_{\Gamma} \nabla_{\Sigma} \Psi + S - Q + \Theta \Pi. \tag{22.3}$$

Eqs. (22.2) state that $\{S_{\Gamma\Sigma} - Q_{\Gamma\Sigma}\}$ admits $\{U_\Gamma\}$ as an eigenvector with the associate eigenvalue being μ. If μ is a simple eigenvalue, as is almost invariably the case owing to the fact that $\{U_\Gamma\}$ is timelike, we obtain a representation for $\{S_{\Gamma\Sigma}\}$, namely

$$S_{\Gamma\Sigma} = Q_{\Gamma\Sigma} + \mu U_\Gamma U_\Sigma + \sigma_{\Gamma\Sigma}, \tag{22.4}$$

where $\{\sigma_{\Gamma\Sigma}\}$ is a symmetric tensor field that admits $\{U_\Gamma\}$ as a null vector. This representation gives $S = Q + \mu + \sigma_\Gamma^\Gamma$, on transvection with $\{a_{\Gamma\Sigma}\}$. When we substitute this result back into (22.3), the following basic equation is obtained for the determination of the function Ψ:

$$\Theta(a^{\Gamma\Sigma} \nabla_\Gamma \nabla_\Sigma \Psi + \Pi) + \sigma_\Gamma^\Gamma = 0. \tag{22.5}$$

23. More about $\{S_{\Gamma\Sigma}\}$

We have fairly exact information concerning the U's, namely, the relations given by (15.3),

$$\begin{gathered} U_\Gamma U^\Gamma = 1, \quad \nabla_\Gamma U_\Sigma = U_\Gamma \partial_\Sigma \Psi, \quad U^\Gamma \partial_\Gamma \Psi = \dot{\Psi} = 0, \\ U^\Gamma \nabla_\Gamma U_\Sigma = \dot{U}_\Sigma = \partial_\Sigma \Psi, \quad \nabla_\Gamma U^\Gamma = 0, \end{gathered} \tag{23.1}$$

while the S's must satisfy the basic existence conditions (20.1),

$$\nabla_\Gamma S_\Sigma^\Gamma = F_\Sigma; \tag{23.2}$$

the latter equations being equivalent to the system

$$\nabla_\Gamma (S_\Sigma^\Gamma - Q_\Sigma^\Gamma) = 0 \tag{23.3}$$

by (20.16). It thus follows, on rewriting (22.4) in the equivalent form

$$S_\Sigma^\Gamma - Q_\Sigma^\Gamma = \mu U_\Sigma U^\Gamma + \sigma_\Sigma^\Gamma, \tag{23.4}$$

that significantly more information can be obtained concerning the detailed structure of the quantities μ and $\{\sigma_\Sigma^\Gamma\}$ that comprise $\{S_{\Gamma\Sigma}\}$.

We first substitute (23.4) into (23.3) to obtain

$$\dot{\mu} U_\Sigma + \mu \dot{U}_\Sigma + \mu U_\Sigma \nabla_\Gamma U^\Gamma + \nabla_\Gamma \sigma_\Sigma^\Gamma = 0. \tag{23.5}$$

An insertion of the results given by (23.1) then gives us

$$\dot{\mu} U_\Sigma + \mu \partial_\Sigma \Psi + \nabla_\Gamma \sigma_\Sigma^\Gamma = 0. \tag{23.6}$$

Transvection of (23.6) with $\{U^\Sigma\}$ and use of (23.1) affords us the relation

$$\dot{\mu} + U^\Sigma \nabla_\Gamma \sigma_\Sigma^\Gamma = 0. \tag{23.7}$$

We now make use of the fact that $\{\sigma_\Sigma^\Gamma\}$ admits $\{U^\Gamma\}$ as a null vector:

$$\sigma_\Sigma^\Gamma U^\Sigma = 0. \tag{23.8}$$

Covariant differentiation of these relations gives

$$0 = \nabla_\Gamma(\sigma^\Gamma_\Sigma U^\Sigma) = U^\Sigma \nabla_\Gamma \sigma^\Gamma_\Sigma + \sigma^\Gamma_\Sigma \nabla_\Gamma U^\Sigma ,$$

and hence, by (23.1), we obtain

$$U^\Sigma \nabla_\Gamma \sigma^\Gamma_\Sigma = -\sigma^\Gamma_\Sigma a^{\Sigma\Lambda} \partial_\Lambda \Psi U_\Gamma = 0 . \tag{23.9}$$

A substitution of (23.9) into (23.7) thus leads to the relation

$$\dot{\mu} = U^\Gamma \partial_\Gamma \mu = 0 . \tag{23.10}$$

The eigenvalue μ is thus a constant on any trajectory of the $\{U^\Gamma\}$ field – it is constant on any solution of $\mathrm{d}u^\Gamma/\mathrm{d}q = U^\Gamma(u^\Sigma)$ – and hence any observer used to establish the geometric equilibrium will always measure the same value of μ as his proper time progresses.

The Eqs. (23.6) now take the simple form

$$\mu \partial_\Sigma \Psi + \nabla_\Gamma \sigma^\Gamma_\Sigma = 0 . \tag{23.11}$$

If we add and subtract the quantity $\partial_\Sigma(\mu\Psi)$, the result can be written in the equivalent form

$$\nabla_\Gamma(\sigma^\Gamma_\Sigma + \mu \Psi \delta^\Gamma_\Sigma) = \Psi \partial_\Sigma \mu . \tag{23.12}$$

Introducing the quantities $\{P^\Gamma_\Sigma\}$ by

$$\sigma^\Gamma_\Sigma = P^\Gamma_\Sigma - \mu \Psi \delta^\Gamma_\Sigma , \qquad P_{[\Gamma\Sigma]} = 0 , \tag{23.13}$$

(23.12) reduces to the system

$$\nabla_\Gamma P^\Gamma_\Sigma = \Psi \partial_\Sigma \mu . \tag{23.14}$$

Since $\{\sigma^\Gamma_\Sigma\}$ must satisfy the conditions (23.8), (23.13) yields

$$P^\Gamma_\Sigma U^\Sigma = \mu \Psi U^\Gamma , \tag{23.15}$$

and we accordingly have

$$P^\Gamma_\Sigma = \mu \Psi U_\Sigma U^\Gamma + W^\Gamma_\Sigma \tag{23.16}$$

where

$$W^\Gamma_\Sigma U^\Sigma = 0 , \qquad W_{[\Gamma\Sigma]} = 0 . \tag{23.17}$$

A combination of (23.14) through (23.17) and use of (23.1) then gives us

$$\nabla_\Gamma W^\Gamma_\Sigma = \mu \Psi \partial_\Sigma (\ln \mu - \Psi) . \tag{23.18}$$

The Eqs. (23.17) and (23.18) are the determining equations for $\{\mathrm{W}_{\Gamma\Sigma}\}$.

Summarizing the results obtained thus far, we have

$$\sigma_{\Gamma\Sigma} = \mu \Psi (U_\Gamma U_\Sigma - a_{\Gamma\Sigma}) + W_{\Gamma\Sigma} \tag{23.19}$$

while

$$S_{\Gamma\Sigma} = Q_{\Gamma\Sigma} + \mu U_\Gamma U_\Sigma - \mu\Psi(a_{\Gamma\Sigma} - U_\Gamma U_\Sigma) + W_{\Gamma\Sigma}. \tag{23.20}$$

A transvection of (23.19) with $\{a^{\Gamma\Sigma}\}$ gives us

$$\sigma_\Gamma^\Gamma = -2\mu\Psi + W_\Gamma^\Gamma,$$

and hence the basic Eq. (22.5) reads

$$\Theta(a^{\Gamma\Sigma}\nabla_\Gamma\nabla_\Sigma\Psi + \Pi) - 2\mu\Psi + W_\Gamma^\Gamma = 0. \tag{23.21}$$

The reader must bear in mind that we have disclaimed knowledge of the momentum-energy tensor, and hence $\{S_{\Gamma\Sigma}\}$ is not known *a priori*; in fact, it is exactly this lack of knowledge that forces us to take the above circuitous course to glean information concerning $\{S_{\Gamma\Sigma}\}$. In general, an arbitrary specification of the momentum-energy tensor will not satisfy the conditions of geometric equilibrium on the discontinuity hypersurface $\mathscr{S}$. It is, however, enlightening to look for a concrete realization in which the above relations obtain. An obvious candidate is the external enveloping environment that was obtained in Chapter I.

In the case of the conformally homogeneous external environment, the salient results have already been established in Section 21:

$$\bar{S}_{\Gamma\Sigma} = -\kappa^{-1} J a_{\Gamma\Sigma} \tag{23.22}$$

where J is a function defined on $\mathscr{S}$ by the relations

$$J = [\![\partial_C\partial_D\psi]\!] N^C N^D. \tag{23.23}$$

Since the momentum-energy tensor for conformally homogeneous models is specified, the quantities $\bar{\chi}$ and $\{\bar{F}_\Sigma\}$ are thus given explicitly by the basic existence conditions (20.1) and (20.2); and (23.22) gives

$$\bar{F}_\Sigma = -\kappa^{-1}\nabla_\Sigma J, \quad \bar{\chi} = \kappa^{-1} J b_\Gamma^\Gamma. \tag{23.24}$$

When (23.22) is compared with (23.20), (23.4), (23.15) and (23.16), we obtain

$$\bar{\sigma}_{\Gamma\Sigma} = \bar{\mu}\Psi(U_\Gamma U_\Sigma - a_{\Gamma\Sigma}) + \bar{W}_{\Gamma\Sigma}, \quad \bar{\mu} = -\kappa^{-1} J, \tag{23.25}$$

$$\bar{W}_{\Gamma\Sigma} = \bar{\mu}(1 + \Psi)(a_{\Gamma\Sigma} - U_\Gamma U_\Sigma) - \bar{Q}_{\Gamma\Sigma}. \tag{23.26}$$

The external enveloping environment obtained in Chapter I is thus consistent with the representations obtained in our analysis of the cases where there is no knowledge concerning the nature of the momentum-energy tensor. The external environment obtained in the previous Chapter therefore provides a natural context in which we may apply the present theory of galactic isopleths. If this were not the case, this Tract would never have been written. At this point we record the following result for future reference.

$$\bar{W}_\Gamma^\Gamma = 2\bar{\mu}(1 + \Psi) - \bar{Q}_\Gamma^\Gamma. \tag{23.27}$$

The reader should note that the perfect fluid momentum-energy tensor, as an exact specification of the momentum-energy tensor, does not satisfy the conditions given above. This follows from (21.1) since the jump in this case gives the surface tensor $\{S^*_{\Gamma\Sigma}\}$ of rank one, while (23.20) demands that $\{S_{\Gamma\Sigma}\}$ have rank greater than one.

24. The Isoplethic Condition and the Perfect Fluid Approximation

The physical identification of an isoplethic surface given in Section 21 was based on the assumption that the perfect fluid approximation was valid to within second order terms. We now give a precise definition of an isoplethic surface as it will be used hereafter. *A hypersurface $\mathcal{S}$ in $\mathcal{E}$ is an isoplethic surface in a state of geometric equilibrium if and only if the eigenvalue, μ, of $\{S_{\Gamma\Sigma}-Q_{\Gamma\Sigma}\}$ that is associated with the eigenvector $\{U_\Gamma\}$ is constant in value.*

We have already seen that $\dot{\mu}=0$ as a consequence of the conditions of geometric equilibrium, and hence the isoplethic condition

$$\partial_\Sigma \mu = 0 \tag{24.1}$$

is consistent with geometric equilibrium and is equivalent to the statement that μ has a constant value on some two-dimensional subspace of the hypersurface $\mathcal{S}$ that is everywhere orthogonal to the congruence generated by $\{U_\Gamma\}$[31]. The reader should note that μ is an eigenvalue of $\{S_{\Gamma\Sigma}-Q_{\Gamma\Sigma}\}$, not of just $\{S_{\Gamma\Sigma}\}$ by itself. Accordingly, since $\{Q_{\Gamma\Sigma}\}$ is the shear potential of the shear vector $\{F_\Sigma\}$, the isoplethic condition refers only to the reduced jump strength tensor $\{S_{\Gamma\Sigma}-Q_{\Gamma\Sigma}\}$. The isoplethic condition thus involves the normal force structure on $\mathcal{S}$ in $\mathcal{E}$ and may accordingly be expected to yield conditions whereby the normal force structure is adjusted in such a fashion that the hypersurface $\mathcal{S}$ has the isoplethic property when it defines a state of geometric equilibrium.

It is now both useful and instructive to compare the isoplethic condition (24.1) with the definition of a physical isopleth obtained in Section 21 under the assumption that the perfect fluid model holds in first approximation[32]. The condition in question is

$$S^*_{\Gamma\Sigma} = -\bar{\rho}\, U_\Gamma U_\Sigma \tag{24.2}$$

where $\bar{\rho}$ is a positive constant that is identified with a constant value of the mean mass density of a galaxy above a background mean mass

[31] This conclusion follows from the fact that $\{U_\Gamma\}$ is tangent to a normal congruence in $\mathcal{S}^*$.

[32] In the first version of the theory, (24.2) was used as a motivation for (24.1). It now seems preferable to give an exact definition of the isoplethic condition and then to use physical approximations to obtain the identification of the quantity μ, rather than the other way around.

density ρ_0. A comparison of Eqs. (24.2) and (23.20) shows that we may tentatively set

$$\mu = -\bar{\rho}\,. \tag{24.3}$$

This identification is, by (23.20), equivalent to the assumption that $\{Q_{\Gamma\Sigma}\}$, $\{W_{\Gamma\Sigma}\}$, and Ψ are quantities of second order[33].

The external environment provided by the conformally homogeneous models lends significant weight to the identification (24.3). We have already seen that the external environment is consistent with the form obtained for $\{S_{\Gamma\Sigma}\}$ in the theory. Now, rather than rely on an approximation, we go back to Section 21 and assume that the conformal factor vanishes exterior to $\mathscr{S}$. In this instance, (21.10) gives us $J = -\kappa[\![{}'\rho]\!]$, while the second of (23.25) gives $\mu = -\kappa^{-1}[\![J]\!]$. In this instance, we thus have the exact result

$$\mu = [\![{}'\rho]\!]\,. \tag{24.4}$$

If ${}'\rho$ has the value ρ_0 just exterior to $\mathscr{S}$, while the value of ${}'\rho$ just interior to $\mathscr{S}$ is $\rho_0 + \bar{\rho}$, the definition of $[\![\cdot]\!]$ gives

$$[\![{}'\rho]\!] = -\bar{\rho}\,. \tag{24.5}$$

A combination of (24.4) and (24.5) thus shows that the external environment with the conformal factor vanishing outside of $\mathscr{S}$ gives the identification (24.3) as a rigorous result that does not depend on approximation. This may be viewed as saying that the perfect fluid approximation in the case considered here for the external environment gives an exact result with respect to the evaluation and identification of μ.

25. The Trace Condition and the Discretization Equation

The basic equation for the determination of Ψ now reads

$$\Theta(a^{\Gamma\Sigma}\nabla_\Gamma\nabla_\Sigma\Psi + \Pi) - 2\mu\Psi + W^\Gamma_\Gamma = 0\,, \tag{25.1}$$

where μ, Θ, and Π are constants and $\{W_{\Gamma\Sigma}\}$ satisfies the conditions

$$W^\Gamma_\Sigma U^\Sigma = 0\,, \qquad W_{[\Gamma\Sigma]} = 0\,, \qquad \nabla_\Gamma W^\Gamma_\Sigma = -\tfrac{1}{2}\partial_\Sigma(\mu\Psi^2)\,. \tag{25.2}$$

Only eight of the nine equations in the system (25.2) are independent, however, for transvection of $\frac{1}{2}\partial_\Sigma(\mu\Psi^2) = \mu\Psi\partial_\Sigma\Psi$ with $\{U^\Sigma\}$ and

---

[33] If we proceeded as in the previous version of the theory referred to in the above footnote, the order of magnitude argument for the identification (24.3) to demand that μ = constant suffers from the unresolvable weakness of the lack of knowledge concerning the Q's, the W's and Ψ. As such, the argument was sloppy, and unnecessarily so. Granted, we have achieved μ = constant by the pontificate of definition, but the definition of the isoplethic condition is a precise statement within the theory and does not wrest itself from mixtures of approximations. An identification, on the other hand, lends itself in a natural fashion to an approximation argument since there one deals with the principal ingredients in establishing such identifications.

$\dot{\Psi} = 0$ gives $(\nabla_\Gamma W^\Gamma_\Sigma) U^\Sigma = 0$, while the same result is implied by $0 = \nabla_\Gamma(W^\Gamma_\Sigma U^\Sigma)$, $\nabla_\Gamma U_\Sigma = U_\Sigma \partial_\Gamma \Psi$ and $W^\Gamma_\Sigma U^\Sigma = 0$. Thus, unless one more condition is specified, the tensor $\{W_{\Gamma\Sigma}\}$ can not be fully determined.

The following condition, which we refer to as the *trace condition* will be assumed:

$$W^\Gamma_\Gamma = \Theta k \,, \qquad \partial_\Sigma k = 0 \,. \tag{25.3}$$

The algebraic interpretation of this assumption is straightforward, for the first of (25.2) gives us

$$\begin{gathered} W_{\Gamma\Sigma} = A X_\Gamma X_\Sigma + B V_\Gamma V_\Sigma \,, \qquad X_\Gamma U^\Gamma = V_\Gamma U^\Gamma = 0 \,, \\ X_\Gamma X^\Gamma = V_\Gamma V^\Gamma = -1 \,. \end{gathered} \tag{25.4}$$

Hence the trace condition yields

$$A + B = -\Theta k \,. \tag{25.5}$$

Since we shall not actually specify the value of the constant k, the constraint (25.3) is a mild one[34]. Even with this assumption, a significantly general structure still remains, for (23.20) gives

$$S_{\Gamma\Sigma} = Q_{\Gamma\Sigma} + \mu U_\Gamma U_\Sigma - \mu \Psi (a_{\Gamma\Sigma} - U_\Gamma U_\Sigma) + W_{\Gamma\Sigma}$$

and the shear potential tensor $\{Q_{\Gamma\Sigma}\}$ still remains at our disposal; it being arbitrary to within the satisfaction of the conditions

$$\nabla_\Gamma Q^\Gamma_\Sigma = F_\Sigma \,.$$

In the case of the general external environment, we have seen that the isoplethic condition gives us

$$\nabla_\Gamma Q^\Gamma_\Sigma = F_\Sigma = 0$$

since J is constant on $\mathscr{S}$. Eq. (23.26) and the trace condition give

$$Q^\Gamma_\Gamma = 2\mu(1 + \Psi) - k\Theta \,.$$

The isoplethic condition and the trace condition thus requires the external environment problem to admit a divergence-free shear potential

[34] The trace condition does not actually have to be assumed for the validity of most of the results given in the next Chapter, at least not for those that derive from the linearized version of the discretization equation. This follows from the linearized version of (25.1) wherein the solution can be represented as the sum of a general solution of the homogeneous equation and a particular solution of the inhomogeneous equation with the inhomogeneous term $-(\Pi + W^\Gamma_\Gamma)$. The trace condition is explicitly included, however, for we wish to make as much of the argument as possible without the assumption of linearization and for the simple fact that the analysis is considerably simplified with the inclusion of (25.3) without undue loss of generality. The reader should carefully note and bear in mind throughout the reading of the next Chapter that the trace condition is not the source of the results we derive and that its elimination will not eliminate the results we obtain. A careful scrutiny will show that it is the isoplethic condition that gives rise to the results.

tensor whose trace is uniquely determined by Ψ and the various constants. The trace of the shear potential tensor in this case is thus constant only in the case where Ψ is constant.

We now have a reasonably simple and straightforward system of equations for the determination of the function Ψ: a combination of the third equation of the system (23.1) with (25.1) and (25.3) gives

$$\Theta(a^{\Gamma\Sigma} \nabla_\Gamma \nabla_\Sigma \Psi + C) = 2\mu\Psi \tag{25.6}$$

and

$$\dot{\Psi} = U^\Gamma \partial_\Gamma \Psi = 0, \quad \partial_\Sigma \mu = 0, \tag{25.7}$$

where the quantity C is a constant whose value is given by

$$C = k + \Pi. \tag{25.8}$$

In view of the results that will be established in the following Chapter, we refer to (25.6) as the *discretization equation.* This equation and its solution, together with the attendant structure of $\mathscr{S}$ is the subject of the next Chapter. The solution of (25.6) is required on the three-dimensional metric space $\mathscr{S}$ and the conditions of geometric equilibrium demand that Ψ be a globally C^2 function on $\mathscr{S}^*$[35].

A brief summary of our findings now seems in order. An isoplethic surface, which is physically defined in terms of the support hypersurface of jump discontinuities in the momentum-energy tensor that satisfy the isoplethic and trace conditions, is in a state of geometric equilibrium if and only if Ψ is a solution of the discretization Eq. (25.5) and the side conditions (25.7). This function Ψ then serves to determine a majority of the properties of the jump strengths, as is easily seen from (23.20), (23.17), (23.18) and (20.2). It is also evident that Ψ serves to determine a significant number of properties of $\mathscr{S}$ since it is the basic defining function of the irrotational isometry that $\mathscr{S}^*$ is required to admit under the conditions of geometric equilibrium. We have also shown that all of the results obtained in this Chapter are consistent with the properties of conformally homogeneous cosmological models when there is a jump discontinuity in the second derivatives of the conformal coefficient, and hence our results are consistent with the external environment. Further, the form of the jump strengths in the momentum-energy tensor given by (23.20) are also consistent with the form of the macroscopic momentum-energy tensor obtained from considerations of the internal consistency environment, since the perfect fluid assumption would have to be augmented by a commutator tensor in order to obtain the structure given by (23.20).

---

[35] If Ψ were not globally C^2 on $\mathscr{S}^*$, the irrotational isometry defined by $\{Y^\Gamma\}$ would rip the isopleth apart or would lead to a situation in which the isopleth would be folded upon itself.

CHAPTER III

Galactic Morphology and Scale

An exact theory of the geometric equilibrium of physical isopleths has been obtained under satisfaction of the isoplethic condition, (24.1), the trace condition, (25.3), and the assumption that the isopleth may be viewed as a timelike hypersurface $\mathscr{S}$ that carries a jump discontinuity in the momentum-energy tensor. The culmination of these considerations was the discretization equation, (25.6), for the determination of the scalar function Ψ that gives the modulus of the vector field that generates the irrotational isometry of $\mathscr{S}$ that is required by the conditions of geometric equilibrium. If the results of the previous Chapter are examined, it is seen that a significant portion of the resulting structure of a physical isopleth in a state of geometric equilibrium is determined by precisely the function Ψ. The properties of the function Ψ, as determined by the solutions of the discretization equations are obtained in this Chapter. Before this is done, however, we implement conditions that relate the general theory of equilibrium isopleths to galactic structures. This is done, in effect, by the requirement that a galactic isoplethic surface admit time-slices that are homotopic to the surface of an oblate spheroid in three-dimensional Euclidean space.

A. The Underlying Geometry of $\mathscr{S}^*$ and the Equations of the Isopleth

The coefficient tensor, $\{a_{\Gamma\Sigma}\}$, of the first fundamental form of $\mathscr{S}$ is a basic element in the discretization Eq. (25.6), as is seen from the identity

$$a^{\Gamma\Sigma} \nabla_\Gamma \nabla_\Sigma \Psi = a^{-1} \partial_\Gamma (a\, a^{\Gamma\Sigma} \partial_\Sigma \Psi), \qquad a = \sqrt{|\det(a_{\Gamma\Sigma})|}\,.$$

Further, when a $*$-coordinate system is introduced in the three-dimensional metric space $\mathscr{S}^*$, which is the metric space of the intrinsic geometry of $\mathscr{S}$, we saw in Section 25 that $a_{00} \overset{*}{=} \exp(-2\Psi)$, and hence the discretization equation is nonlinear. It thus follows that if we are to obtain an explicit form for the discretization equation, we must evaluate the first fundamental form. As might be expected, such an evaluation leads to significant information concerning the structure of the hypersurface $\mathscr{S}$. Although the information obtained in this way must be interpreted in something of a round about fashion owing to our lack of knowledge concerning the detailed metric structure of the enveloping space $\mathscr{E}$, a direct means is obtained for the explicit description of the isoplethic surface in terms of Euclidean constructs. This will be accomplished by implementing the requirement that the time-wise cross-sections of a galactic world tube must be homotopic to the surface of an oblate spheroid in three-dimensional space. Such a requirement is obviously necessary, for the general theory of equilibrium isopleths obtained in the previous Chapter is applicable to a wide class of time-like hypersurfaces that need have nothing whatever to do with galactic structures.

26. Interpretation of the Ψ Function

We shall first investigate the significance and interpretation of the Ψ function in view of the relation $a_{00} \overset{*}{=} \exp(-2\Psi)$. With a $*$-coordinate system introduced in Section 15, we have

$$U^{\Gamma} \overset{*}{=} \delta_0^{\Gamma} \exp(\Psi), \tag{26.1}$$

while the third of the system of Eqs. (15.3) yields

$$0 = \dot{\Psi} = U^{\Gamma} \partial_{\Gamma} \Psi \overset{*}{=} \exp(\Psi) \partial_0 \Psi,$$

and hence we have

$$\partial_0 \Psi \overset{*}{=} 0. \tag{26.2}$$

Further, from (15.4) and (26.1), we obtain

$$du^{\Gamma}/dq = U^{\Gamma} \overset{*}{=} \delta_0^{\Gamma} \exp(\Psi), \quad q = p \exp(\Psi), \tag{26.3}$$

where q has been identified with the proper time of the observers attached to the isoplethic surface in order to describe the condition of geometric equilibrium. We thus see that Ψ describes variations among the proper clock rates of the observers on the isopleth, and as such, the Ψ function describes the possible states of temporal inhomogeneity of an isoplethic surface in geometric equilibrium. Accordingly, since a temporal inhomogeneity in general relativity is accompanied by a spatial inhomogeneity, it is to be expected that Ψ also serves to describe possible states of spatial inhomogeneity. The assertion is similar to and reminiscent of that argued in Part A of Chapter I.

Let us, for the moment, consider the case in which the space $\mathscr{E}$ admits spherically symmetric time slices. The time slices of the hypersurface $\mathscr{S}$ must then be surfaces of three-spheres if $\mathscr{S}$ is to carry jump discontinuities in the momentum-energy tensor, for any other configuration would violate the assumed spherical symmetry. In this case, the most general form that the jump strengths of the momentum-energy tensor can assume is given by[1]

$$\hat{S}^{\Sigma}_{\Gamma} = L^{-2} a_{00}(A-B)L_{\Gamma}L^{\Sigma} + B\delta^{\Sigma}_{\Gamma}, \qquad L_{\Sigma} = L\delta^{0}_{\Sigma}, \tag{26.4}$$

where the quantities A, B and L are functions of u^0 only. In the particularly simple case where $Q_{\Gamma\Sigma} = 0$[2], (23.20) gives

$$S = \mu(1-2\Psi) + W^{\Gamma}_{\Gamma} = \mu(1-2\Psi) + k\Theta, \tag{26.5}$$

while (26.4) gives

$$\hat{S} = 2B + A. \tag{26.6}$$

A comparison of (26.5) and (26.6) and the fact that k and μ are constants while Ψ is u^0-independent in a $*$-coordinate system and A and B are functions of u^0 only, shows that we can only have Ψ = constant in this case. Conversely, for $\Psi \neq$ constant, the isopleth must exhibit deviations from spherical symmetry. We may thus view the Ψ function as describing a state of perturbation from some underlying symmetry. If we use the results established in the first Chapter concerning the internal consistency environment, and note that the argument of spherical symmetry is almost invariably made within the context of the terms which we have identified with the tensor $\{\langle t^{A}_{B}\rangle\}$, the function Ψ serves partially to determine the surface-projected jump strengths of the field $\{C^{B}_{A}\}$ of commutators. From the standpoint of the external enveloping environment, the Ψ function may be identified in terms of a quantity that arises when attempts are made to join the internal and external structural environments across a boundary surface $\mathscr{S}$, and partially reflects the inhomogeneities demanded by the cosmological state.

27. The Subsidiary Space $\mathscr{C}$

Since the symmetries exhibited by a space significantly simplify the analysis of its properties and structure, it is useful to introduce the concept of a subsidiary space that has the appropriate symmetries and that may be used as an underlying space for the representation and analysis of the asymmetries and inhomogeneities of the actual space under study. We have already seen that the Ψ function describes deviations from states that have basic symmetries, and that the space $\mathscr{S}^*$ is effectively characterized by the single scalar function Ψ. If we combine these facts with the

[1] See Appendix B.

[2] The vanishing of the shear potential tensor is almost tantamount to the assumption of spherical symmetry, for a nonvanishing shear force over a two-dimensional spherical surface will always give rise to a violation of spherical symmetry.

results established in Chapter I concerning the external enveloping environment, we are naturally led to consider a distinct, three-dimensional, hyperbolic-normal metric space $\mathscr{C}$ with metric tensor $\{\bar{a}_{\Gamma\Sigma}\}$, and to require that $\mathscr{C}$ and $\mathscr{S}^*$ be conformally related with parameter λ:

$$a_{\Gamma\Sigma} = \exp(2\lambda)\bar{a}_{\Gamma\Sigma}. \tag{27.1}$$

Certain notational conventions are required in order to avoid an otherwise unwieldy notation. The process of covariant differentiation in $\mathscr{C}$ will be denoted by $\bar{\nabla}$, while ∇ will continue to denote covariant differentiation in $\mathscr{S}^*$. The raising and lowering of indices in $\mathscr{C}$ is accomplished by $\{\bar{a}_{\Gamma\Sigma}\}$ and $\{\bar{a}^{\Gamma\Sigma}\}$, a dot being placed in the original position of the index to avoid confusion with the raising or lowering of indices by means of $\{a_{\Gamma\Sigma}\}$ and $\{a^{\Gamma\Sigma}\}$. A geometric quantity in $\mathscr{C}$ will be denoted by a superimposed bar, as in (27.1) and in the notation for covariant differentiation in $\mathscr{C}$. Thus, the Ricci tensor and the Christoffel symbols of the second kind of $\mathscr{C}$ are denoted by $\{\bar{R}_{\Gamma\Sigma}\}$ and $\{\bar{\Gamma}^{\Gamma}_{\Sigma\Lambda}\}$, respectively.

We assume that the coordinate patches and coordinate functions of $\mathscr{S}^*$ and $\mathscr{C}$ correspond (affine path correspondence). We may thus define a vector field $\{\bar{U}^{\Gamma}\}$ on $\mathscr{C}$ by the relations

$$\bar{U}^{\Gamma} = \exp(\lambda)U^{\Gamma}. \tag{27.2}$$

The vectors with components $\{\bar{U}_{\dot{\Gamma}}\}$ and $\{\bar{U}_{\Gamma}\}$ are then defined by

$$\bar{U}_{\Gamma} = a_{\Gamma\Sigma}\bar{U}^{\Sigma}, \qquad \bar{U}_{\dot{\Gamma}} = \bar{a}_{\Gamma\Sigma}\bar{U}^{\Sigma} = \exp(-2\lambda)\bar{U}_{\Gamma} = \exp(-\lambda)U_{\Gamma}. \tag{27.3}$$

Thus, $\{\bar{U}^{\Gamma}\}$ is a unit, timelike vector field on $\mathscr{C}$:

$$\bar{U}^{\Gamma}\bar{U}^{\Sigma}\bar{a}_{\Gamma\Sigma} = \bar{U}^{\Gamma}\bar{U}_{\dot{\Gamma}} = U^{\Gamma}U_{\Gamma} = 1. \tag{27.4}$$

The conformal relation between $\mathscr{S}^*$ and $\mathscr{C}$ allows us to use the well known properties of such spaces to write all of the previous results for $\mathscr{S}^*$ in terms of the corresponding results in $\mathscr{C}$. The admission of a time oriented irrotational isometry by $\mathscr{S}^*$ then conveys to $\mathscr{C}$ certain properties that are delineated in Appendix C, and will be quoted without further note.

From $U^{\Gamma}\partial_{\Gamma}\Psi = 0$ (see (25.7)), we know that Ψ is constant on a trajectory of the $\{U^{\Gamma}\}$ field in $\mathscr{S}^*$. It is thus natural to expect that the geometric perturbations of $\mathscr{S}^*$ that result from Ψ are constant on the trajectories of the $\{U^{\Gamma}\}$ field; indeed, otherwise the conditions of geometric equilibrium would be violated. Thus, since the geometry of $\mathscr{C}$ is to be without such geometric perturbations, it follows from (27.1) that λ must also be constant on the trajectories of the $\{U^{\Gamma}\}$ field:

$$U^{\Gamma}\partial_{\Gamma}\lambda = 0. \tag{27.5}$$

With this condition, we obtain (see Appendix C)

$$\bar{U}^{\Gamma} \bar{R}_{\Gamma\Sigma} = \bar{U}^{\cdot}_{\Sigma} \bar{a}^{\Gamma\Lambda} \bar{V}_{\Gamma} \bar{V}_{\Lambda} \omega \tag{27.6}$$

and

$$\bar{U}^{\Gamma} \bar{V}_{\Gamma} \bar{U}^{\cdot}_{\Sigma} = \partial_{\Sigma} \omega \tag{27.7}$$

in $\mathscr{C}$, where ω is the "λ-differentia" defined by

$$\omega = \lambda + \Psi . \tag{27.8}$$

The discretization Eq. (25.6), when carried over to $\mathscr{C}$, assumes the form

$$(\bar{V}_{\Gamma} + \partial_{\Gamma}\omega - \partial_{\Gamma}\Psi)\partial^{\Gamma}\Psi + (C - 2\mu\Psi/\Theta)\exp(2\omega - 2\Psi) = 0 , \tag{27.9}$$

while (25.7) and the trace condition become

$$\bar{U}^{\Gamma} \partial_{\Gamma} \Psi = 0 \tag{27.10}$$

$$W_{\Gamma\Sigma} \bar{a}^{\Gamma\Sigma} = \Theta k \exp(2\lambda) , \tag{27.11}$$

respectively (see Appendix C).

28. The Metric Tensor of $\mathscr{C}$

In view of the relations (15.1), (27.2) and (27.8), we have

$$\bar{U}^{\Gamma} = \exp(\lambda) U^{\Gamma} = \exp(\lambda + \Psi) Y^{\Gamma} = \exp(\omega) Y^{\Gamma} . \tag{28.1}$$

It thus follows that when a $*$-coordinate system is introduced in $\mathscr{S}^*$ by the relations (15.5), a corresponding $*$-coordinate system is introduced in $\mathscr{C}$ by the relations

$$\bar{U}^{\Gamma} \overset{*}{=} \exp(\omega)\, \delta^{\Gamma}_{0} . \tag{28.2}$$

Recalling that $U^{\Gamma}\partial_{\Gamma}\lambda = U^{\Gamma}\partial_{\Gamma}\Psi = 0$, and hence that $U^{\Gamma}\partial_{\Gamma}\omega = 0$ by (27.8), it follows that

$$\partial_0 \lambda \overset{*}{=} \partial_0 \Psi \overset{*}{=} \partial_0 \omega \overset{*}{=} 0 \tag{28.3}$$

hold in $\mathscr{C}$. When (27.1) is referred to the corresponding $*$-coordinate systems in $\mathscr{S}^*$ and $\mathscr{C}$, and use is made of (15.6) and (15.7), the metric tensor on $\mathscr{C}$ is seen to satisfy the conditions

$$\partial_0 \bar{a}_{\Gamma\Sigma} \overset{*}{=} 0, \qquad \bar{a}_{00} \overset{*}{=} \exp(-2\omega), \qquad \bar{a}_{0\beta} \overset{*}{=} 0 . \tag{28.4}$$

Eqs. (28.4) provide us with an explicit evaluation of the quantities $\{\bar{a}_{0\Gamma}\}$ and the information that all of the $\bar{a}$'s are u^0-independent in a $*$-coordinate system on $\mathscr{C}$. Thus, since $\mathscr{C}$ is to represent the unperturbed geometry corresponding to $\mathscr{S}^*$, while the isopleths to be identified with $\mathscr{S}^*$ should include the elliptical galaxies in the case of negligible Ψ variation, we take

$$\bar{a}_{11} \overset{*}{=} -r^2(1 - \varepsilon^2 \sin^2 u^1), \qquad \bar{a}_{22} \overset{*}{=} -r^2 \sin^2 u^1 , \qquad \bar{a}_{12} \overset{*}{=} 0 , \tag{28.5}$$

where the quantities r and ε refer to the semimajor axis and the eccentricity, respectively, of an oblate spheroid. This choice is evident from the fact that $\mathrm{d}J^2 = -\bar{a}_{\alpha\beta}\mathrm{d}u^\alpha \mathrm{d}u^\beta$ is the line element on the surface of an oblate spheroid in Euclidean three-dimensional space with surface coordinates u^1 = elliptic colatitude, u^2 = elliptic longitude, r = semimajor axis, and ε = eccentricity. The signs of $\bar{a}_{11}$ and $\bar{a}_{22}$ are determined by the requirement that $\mathscr{C}$ be hyperbolic-normal and the fact that we have $\bar{a}_{0\beta} \overset{*}{=} 0$, that is

$$\bar{a}_{\Gamma\Sigma}\mathrm{d}u^\Gamma \mathrm{d}u^\Sigma \overset{*}{=} \exp(-2\omega)(\mathrm{d}u^0)^2 + \bar{a}_{\alpha\beta}\mathrm{d}u^\alpha \mathrm{d}u^\beta .$$

The choice (28.5) is what converts our previous analysis of general isoplethic surfaces in states of geometric equilibrium to the study of equilibrium isoplethic surfaces of galaxies, for (28.5) renders the time slices of $\mathscr{S}$ homotopic to oblate spheroids, and, as we shall see, provides the global conditions that must be imposed in order to supplement the local conditions obtained in the previous Chapter if we are indeed discussing galaxies. Summarizing, the metric tensor on $\mathscr{C}$ is given in a $*$-coordinate system by

$$\bar{a} \overset{*}{=} \begin{pmatrix} \exp(-2\omega) & 0 & 0 \\ 0 & -r^2(1-\varepsilon^2\sin^2 u^1) & 0 \\ 0 & 0 & -r^2\sin^2 u^1 \end{pmatrix}. \tag{28.6}$$

29. The Embedding Problem

To this point, we have evaluated the quantities $\{\bar{a}_{\Gamma\Sigma}\}$, and hence, by (27.1), the quantities $\{a_{\Gamma\Sigma}\}$ to within the factor $\exp(2\lambda)$ in a $*$-coordinate system:

$$a_{\Gamma\Sigma} \overset{*}{=} \begin{pmatrix} \exp(-2\Psi) & 0 & 0 \\ 0 & -r^2(1-\varepsilon^2\sin^2 u^1)\exp(2\lambda) & 0 \\ 0 & 0 & -r^2\sin^2 u^1\exp(2\lambda) \end{pmatrix}. \tag{29.1}$$

Furthermore, this evaluation has involved no explicit consideration of the metric structure of the enveloping space $\mathscr{E}$, although we know that $\mathscr{E}$ must admit the embedding of $\mathscr{S}$ as a time-like hypersurface. The implications of this embedding condition will now be obtained.

Let us first establish the existence of a four-dimensional, hyperbolic-normal metric space $\bar{\mathscr{E}}$ in which the space $\mathscr{C}$ can be embedded as a time-like hypersurface. We denote the fundamental metric differential form of $\bar{\mathscr{E}}$ by

$$\mathrm{d}\bar{s}^2 = \bar{h}_{AB}\mathrm{d}y^A \mathrm{d}y^B$$

and the parametric equations of a hypersurface in $\bar{\mathscr{E}}$ by $y^A = \bar{f}^A(u^\Gamma)$[3]. The coefficient tensor of the first fundamental form given by (28.2) is then reproduced from the defining equations by

$$\begin{aligned} &\bar{f}^0 \stackrel{*}{=} u^0, \quad \bar{f}^1 \stackrel{*}{=} r\sin u^1 \cos u^2, \\ &\bar{f}^2 \stackrel{*}{=} r\sin u^1 \sin u^2, \quad \bar{f}^3 \stackrel{*}{=} r(1-\varepsilon^2)^{\frac{1}{2}}\cos u^1 \end{aligned} \tag{29.2}$$

provided that

$$\mathrm{d}\bar{s}^2 \stackrel{*}{=} \exp(-2\Omega(y^K))(\mathrm{d}y^0)^2 + \mathrm{d}y^0 M_i(y^K)\mathrm{d}y^i - (\mathrm{d}y^1)^2 - (\mathrm{d}y^2)^2 - (\mathrm{d}y^3)^2 \tag{29.3}$$

and

$$\Omega(\bar{f}^K(u^\Gamma)) \stackrel{*}{=} \omega(u^\Gamma), \qquad M_i(\bar{f}^K(u^\Gamma))\, y^i_\Sigma \stackrel{*}{=} 0\,. \tag{29.4}$$

Since the function $\omega(u^\Gamma)$ is known at this point only in that it satisfies the condition $\partial_0\omega \stackrel{*}{=} 0$, the first of Eqs. (29.4) shows that $\Omega(y^K)$ is likewise arbitrary to within the corresponding restriction $(\partial_A\Omega)\,y^A_0 \stackrel{*}{=} 0$. In view of the first of Eqs. (29.2), however, this restriction amounts to $\partial_0\Omega| \stackrel{*}{=} 0$, where $\Omega|$ denotes the restriction of Ω to $\mathscr{C}$.

It now follows from the form of the line element of $\bar{\mathscr{E}}$ given by (29.3) that the quantities $\{M_i\}$ form the components of a vector field in the three-dimensional Euclidean space obtained from $\bar{\mathscr{E}}$ by the restriction $y^0 =$ constant. Thus, since $y^i_0 \stackrel{*}{=} 0$, by (29.2), the second equation in (29.4) shows that the vector field with components $\{M_i(\bar{f}^K(u^\Gamma))\}$ is orthogonal to the two-dimensional surface obtained from the intersection of $\mathscr{C}$ with the hypersurface $y^0 =$ constant. Hence, for the special case $M_i(\bar{f}^K) \stackrel{*}{=} 0$, the resulting line element shows that $\bar{\mathscr{E}}$ admits the simultaneous embedding of the two-parameter family of spaces $\mathscr{C}$ that obtains by letting r and ε range through all admissible values. Consequently, for $M_i(\bar{f}^K) \stackrel{*}{=} 0$, the space $\bar{\mathscr{E}}$ is a universal immersion space for all $\mathscr{C}$'s. For $M_i(\bar{f}^K) \neq 0$, $\bar{\mathscr{E}}$ is a universal immersion space for all $\mathscr{C}$'s with fixed ε provided the M's are tangent to the curves

$$(y^3)^2 = A\{(y^1)^2 + (y^2)^2\}^{(1-\varepsilon)/2}, \qquad A > 0\,.$$

Now, the space $\bar{\mathscr{E}}$ has not been connected with the Einstein theory; in these considerations it has been obtained solely as a hyperbolic-normal metric space in which $\mathscr{C}$ can be embedded as a timelike hypersurface. However, in view of the fact that $\mathscr{C}$ represents the unperturbed geometry underlying $\mathscr{S}$, it would be natural to expect that $\bar{\mathscr{E}}$ would represent the unperturbed geometry of $\mathscr{E}$ in some appropriate sense. Such a result is also to be expected in view of the results obtained in Part A of Chapter I. In the general case ($M_i \neq 0$), the form given by (29.3) is that obtained by

[3] The coordinate patches and coordinate functions of $\mathscr{C}$ and $\mathscr{S}^*$ coincide and hence it is not necessary to distinguish between $\{u^\Gamma\}$ and $\{\bar{u}^\Gamma\}$ in a $*$-coordinate system.

Kirkwood[4] by a direct representation of the experimentally verifiable facts of gravitational phenomena. For the case $M_i = 0$, the line element given by (29.3) is that which results from the Einstein theory in first approximation[5] if we exercise the freedom to choose Ω and take

$$\Omega = -\tfrac{1}{2}\ln(1 + 2\Phi/c^2). \tag{29.5}$$

Here, Φ denotes the Newtonian gravitational potential and c is the speed of light. This approximation is also afforded by the form obtained when $M_i \neq 0$[6]. According to the known smallness of relativistic gravitational effects, at least as far as the metric geometry is concerned, the line element given by (29.3) is an excellent first approximation.

We have thus established the following rather surprising result. The metric properties of the subsidiary space $\mathscr{C}$, when embedded as a volume-filling family of timelike hypersurfaces in a hyperbolic-normal, four- dimensional metric space $\bar{\mathscr{E}}$, leads to a determination of the metric structure of $\bar{\mathscr{E}}$ that is adequate to represent the Einstein field in first approximation.

The first of the relations (29.4) and (29.5) give us the quantity ω according to the formula

$$\omega = -\tfrac{1}{2}\ln(1 + 2\Phi|/c^2), \tag{29.6}$$

where $\Phi|$ denotes the restriction of Φ to the hypersurface $\mathscr{C}$. Using this to rewrite (27.6) and (27.7), we have

$$\bar{U}^\Gamma \bar{R}_{\Gamma\Sigma} \stackrel{*}{=} -\bar{U}^{\cdot}_{\Sigma}\bar{a}^{\Gamma\Lambda}\bar{V}_\Gamma\{\partial_\Lambda\Phi|/c^2(1 + 2\Phi|/c^2)\} \tag{29.7}$$

$$\bar{U}^\Gamma \bar{V}_\Gamma \bar{U}^{\cdot}_{\Sigma} = -\partial_\Sigma\Phi|/c^2(1-2\Phi|/c^2). \tag{29.8}$$

Neglecting terms of order c^{-2} (Newtonian approximation), the curvature of $\mathscr{C}$ in the $\{\bar{U}^\Gamma\}$ direction vanishes and the trajectories of the $\{\bar{U}^\Gamma\}$ field are geodesics. Retaining terms of order c^{-2} in (29.8), we see that an effective force $\{-\partial_\Sigma\Phi|\}$ has to be applied in $\mathscr{C}$ in order to generate the trajectories of the $\{\bar{U}^\Gamma\}$ field. This states that the surface $\mathscr{C}$ is maintained in geometric equilibrium in $\bar{\mathscr{E}}$ by forces other than just those of gravitation, and these additional forces exactly cancel the gravitational forces when projected onto $\mathscr{C}$.

Because $\mathscr{S}^*$ is embedded in $\mathscr{E}$, while $\mathscr{C}$ is embedded in $\bar{\mathscr{E}}$, the second system of Eq. (13.3) gives

$$h^{AB} = a^{\Gamma\Sigma}x^A_\Gamma x^B_\Sigma - N^A N^B, \quad \bar{h}^{AB} = \bar{a}^{\Gamma\Sigma}y^A_\Gamma y^B_\Sigma - \bar{N}^A\bar{N}^B, \tag{29.9}$$

while the first of (13.3) yields

$$a_{\Gamma\Sigma} = h_{AB}x^A_\Gamma x^B_\Sigma, \quad \bar{a}_{\Gamma\Sigma} = \bar{h}_{AB}y^A_\Gamma y^B_\Sigma. \tag{29.10}$$

---

[4] Kirkwood, R. L.: Lorentz Invariance in a Gravitational Field. The RAND Corporation, RM-3146-RC (1963).

[5] Tolman, R. C.: *Relativity, Thermodynamics and Cosmology*. Oxford: Clarendon Press 1934.

[6] Kirkwood, R. L.: Lorentz Invariance in a Gravitational Field. The RAND Corporation, RM-3146-RC (1963).

It thus follows from the conformal relation (27.1) between $\mathscr{S}^*$ and $\mathscr{C}$ that

$$h_{AB} x_\Gamma^A x_\Sigma^B = \exp(2\lambda) \bar{h}_{AB} y_\Gamma^A y_\Sigma^B . \tag{29.11}$$

Noting that all terms on the right-hand side of these relations are known in a $*$-coordinate system, we have six equations that must be satisfied by the quantities $\{x_\Gamma^A\}$ and $\{h_{AB}\}$ on $\mathscr{S}$ in $\mathscr{E}$. Since $\mathscr{S}^*$ and $\mathscr{C}$ have the same coordinate patches and coordinate functions, while (29.9) shows that it is consistent to assume that the differences between $\mathscr{S}^*$ and $\mathscr{C}$ arise from the different properties of the spaces $\mathscr{E}$ and $\bar{\mathscr{E}}$ (i. e., form the different tensors $\{h_{AB}\}$ and $\{\bar{h}_{AB}\}$)[7], we explicitly assume that $\mathscr{S}^*$ and $\mathscr{C}$ have the same parametric equations modulo a group of isotropic translations in $\mathscr{E}$:

$$x_\Gamma^A(u^\Sigma) = y_\Gamma^A(u^\Sigma) . \tag{29.12}$$

Accordingly, since $\{N_A\}$ and $\{\bar{N}_A\}$ satisfy the relations

$$N_A x_\Gamma^A = 0, \qquad \bar{N}_A y_\Gamma^A = 0, \qquad N_A N_B h^{AB} = \bar{N}_A \bar{N}_B \bar{h}^{AB} = -1,$$

we must have

$$\bar{N}_A = K N_A \tag{29.13}$$

for some $K > 0$. A simple calculation based on (29.9), (29.11) and (29.12), and the conformal relation (27.1) then yields

$$h^{AB} = \exp(-2\lambda)(\bar{h}^{AB} + \bar{N}^A \bar{N}^B) - N^A N^B \tag{29.14}$$

on $\mathscr{S}$ in $\mathscr{E}$. We also have, from (29.11), (29.12) and (29.13), that

$$h_{AB} = \exp(2\lambda) \bar{h}_{AB} + B \bar{N}_A \bar{N}_B, \qquad B \neq \exp(2\lambda), \tag{29.15}$$

where the latter inequality follows from the fact that both $\{h_{AB}\}$ and $\{\bar{h}_{AB}\}$ are nonsingular.

30. Spacelike Quadrics

Eqs. (29.3) show that the spacelike surface $\bar{\mathscr{T}}$, defined in $\bar{\mathscr{E}}$ by the equation $y^0 =$ constant, is Euclidean, while (29.2) and (29.3) show that the intersection of $\mathscr{C}$ with $\bar{\mathscr{T}}$ is the surface of an oblate spheroid. On the other hand, the realization (29.15) of the metric tensor of $\mathscr{E}$ on $\mathscr{S}$ shows that the surface $\mathscr{T}$ in $\mathscr{E}$, defined by $x^0 =$ constant, can not be Euclidean if λ is other than a constant. We thus have to obtain a consistent means of interpreting the surface $\mathscr{S} \cap \mathscr{T}$, since this surface is the isoplethic surface of a galaxy in three-dimensional space.

[7] We have disclaimed knowledge of the momentum-energy tensor of $\mathscr{E}$, and hence, if the Einstein field equations hold, we must equivalently disclaim knowledge of the metric tensor of $\mathscr{E}$. The basic purpose of the work undertaken here is to get some sort of hold on the metric tensor of $\mathscr{E}$ in a neighborhood of $\mathscr{S}$ by use of the fact that $\mathscr{S}$ must be embedded in $\mathscr{E}$.

Consider the space $\tilde{\mathscr{E}}$ whose fundamental metric differential form is given by

$$d\tilde{s}^2 = \tilde{h}_{AB}\, dx^A dx^B \stackrel{*}{=} \exp(2\rho)\bar{h}_{AB}\, dx^A dx^B \tag{30.1}$$

for bounded $\rho(x^K)$. A comparison of (29.15) and (30.1) shows that $\tilde{\mathscr{E}}$ admits the simultaneous embedding of all $\mathscr{S}$'s for all possible values of r and ε if we have

$$\rho(f^A(u^\Gamma)) = \lambda(u^\Gamma). \tag{30.2}$$

The space $\tilde{\mathscr{E}}$ is thus a universal immersion space for all $\mathscr{S}$'s obtained from all $\mathscr{C}$'s by conformal relation.

We now confine our attention to quantities referred to $*$-coordinate systems. It is evident from (29.3) that $\{\bar{h}_{ij}\}$ is a negative definite symmetric tensor field on the Euclidean three-dimensional space $\bar{\mathscr{E}} \cap \bar{\mathscr{T}}$. We may therefore define a general quadric $\bar{\mathscr{Q}}$ on $\bar{\mathscr{E}} \cap \bar{\mathscr{T}}$ by the equation

$$\bar{h}_{ij}\bar{X}^i\bar{X}^j \stackrel{*}{=} -1. \tag{30.3}$$

The quantities $\{\bar{X}^i\}$ are the quadric coordinates of $\bar{\mathscr{Q}}$; they generate realizations of $\bar{\mathscr{Q}}$ as two dimensional surfaces in the three-dimensional Euclidean space $\bar{\mathscr{E}} \cap \bar{\mathscr{T}}$ under the substitution

$$\bar{X}^i(y^k) = A^i_j y^j + B^i,$$

where the quantities $\{A^i_j\}$ and $\{B^i\}$ are constants and $\det(A^i_j) > 0$. In particular, the surface of the oblate spheroid $\mathscr{C} \cap \bar{\mathscr{T}}$ is a realization of $\bar{\mathscr{Q}}$ that obtains under the substitution

$$\bar{X}^1 = y^1/r, \quad \bar{X}^2 = y^2/r, \quad \bar{X}^3 = y^3/(r(1-\varepsilon^2)^{\frac{1}{2}}).$$

Similarly, since $\{\tilde{h}_{ij}\}$ is negative definite by (29.3) and (30.1), a general quadric $\tilde{\mathscr{Q}}$ can be defined in $\tilde{\mathscr{E}} \cap \tilde{\mathscr{T}}$ by

$$\tilde{h}_{ij}\tilde{X}^i\tilde{X}^j \stackrel{*}{=} -1, \tag{30.4}$$

where $\tilde{\mathscr{T}}$ is the hypersurface in $\tilde{\mathscr{E}}$ defined by $x^0 \stackrel{*}{=}$ constant.

We now inquire into the relation between the general quadrics $\tilde{\mathscr{Q}}$ and $\bar{\mathscr{Q}}$. Since the $\bar{h}$'s and the $\tilde{h}$'s are related by (30.1), (30.4) can be written in the equivalent form

$$\bar{h}_{ij}(\exp(\rho)\tilde{X}^i)(\exp(\rho)\tilde{X}^j) \stackrel{*}{=} -1. \tag{30.5}$$

It thus follows from a comparison of (30.3) and (30.5) that

$$\tilde{X}^i = \exp(-\rho)\bar{X}^i \tag{30.6}$$

establishes a one-to-one mapping of $\bar{\mathscr{Q}}$ onto $\tilde{\mathscr{Q}}$. Thus, to every realization of $\bar{\mathscr{Q}}$ we can determine a unique corresponding realization of $\tilde{\mathscr{Q}}$, and conversely.

The space $\bar{\mathscr{E}} \cap \bar{\mathscr{T}}$ is a Euclidean three-dimensional space with Cartesian coordinates in the $*$-coordinate system under use; hence, the quadrics in $\bar{\mathscr{E}} \cap \bar{\mathscr{T}}$ are realized as quadric surfaces in the usual sense.

On the other hand, since $\tilde{\mathscr{E}} \cap \tilde{\mathscr{T}}$ is Euclidean if and only if the function ρ is a constant, the question naturally arises as to the meaning of a realization of $\tilde{\mathscr{Q}}$. The obvious answer to this question is to refer the quadric $\bar{\mathscr{Q}}$ to the space $\bar{\mathscr{E}} \cap \bar{\mathscr{T}}$ where quadrics are simple known constructs. This is done uniquely by interpreting the mapping (30.6) as a *quadric point transformation* in $\bar{\mathscr{E}} \cap \bar{\mathscr{T}}$. Under this interpretation, a point on the quadric $\bar{\mathscr{Q}}$ with quadric coordinates $\{\bar{X}^i\}$ is taken into the point in $\bar{\mathscr{E}} \cap \bar{\mathscr{T}}$ with coordinates

$$'\bar{X}^i = \exp(-\rho)\bar{X}^i . \qquad (30.7)$$

The geometric figure in $\bar{\mathscr{E}} \cap \bar{\mathscr{T}}$ obtained by application of the quadric point transformation (30.7) to $\bar{\mathscr{Q}}$ is such that its points satisfy the relation

$$\bar{h}_{ij}\,'\bar{X}^i\,'\bar{X}^j \overset{*}{=} -1 .$$

A comparison of this equation with (30.4) leads to the conclusion that the quadric point transformation (30.7) gives a unique realization of $\tilde{\mathscr{Q}}$ in $\bar{\mathscr{E}} \cap \bar{\mathscr{T}}$ for every realization of $\tilde{\mathscr{Q}}$ in $\tilde{\mathscr{E}} \cap \tilde{\mathscr{T}}$. The uniqueness follows from the uniqueness of the mapping defined by (30.6).

In general, the image of a quadric under a quadric point transformation is no longer a quadric. This is easily seen since the image of $\bar{\mathscr{Q}}$ under the transformation (30.7) is obtained by replacing $\{\bar{X}^i\}$ in (30.3) by $\{'\bar{X}^i\}$ and then express $\{'\bar{X}^i\}$ in terms of $\{\bar{X}^i\}$ by (30.7); the result is

$$\bar{h}_{ij}\bar{X}^i\bar{X}^j \overset{*}{=} -\exp(2\rho) . \qquad (30.8)$$

31. Equations for the Isopleths

An isopleth in three-dimensional space, according to the above constructions, is the intersection $\mathscr{S} \cap \mathscr{T}$. When (30.2) is used, the isopleth is also the realization of $\tilde{\mathscr{Q}}$ in $\tilde{\mathscr{E}} \cap \tilde{\mathscr{T}}$ that corresponds to the quadric figure $\mathscr{C} \cap \bar{\mathscr{T}}$ in $\bar{\mathscr{E}} \cap \bar{\mathscr{T}}$. Now, the image in $\bar{\mathscr{E}} \cap \bar{\mathscr{T}}$ of the quadric figure in $\mathscr{E} \cap \mathscr{T}$ that corresponds to $\mathscr{C} \cap \bar{\mathscr{T}}$ has the equation

$$(y^1/r)^2 + (y^2/r)^2 + (y^3/r)^2/(1-\varepsilon^2) \overset{*}{=} \exp(2\rho) . \qquad (31.1)$$

This follows from (30.8) and the use of the quadric coordinates

$$\bar{X}^1 = y^1/r , \quad \bar{X}^2 = y^2/r , \quad \bar{X}^3 = y^3/r(1-\varepsilon^2)^{\frac{1}{2}} \qquad (31.2)$$

that give the surface of the oblate spheroid $\mathscr{C} \cap \bar{\mathscr{T}}$ by (30.3). With an evaluation of λ in terms of ω and Ψ by (27.8) and replacing ρ by λ in accordance with (30.2), the equations for the isopleth in Euclidean space $\bar{\mathscr{E}} \cap \bar{\mathscr{T}}$ is given by

$$(y^1/r)^2 + (y^2/r)^2 + (y^3/r)^2/(1-\varepsilon^2) = \exp(2\omega - 2\Psi) , \qquad (31.3)$$

where (y^1, y^2, y^3) are Cartesian coordinates. This was the whole purpose of the previous section, for the results obtained there allows us, in effect, to obtain an Euclidean construct inside $\mathscr{S}$ that is such that an external observer would observe (31.3) if he assumed that his observations of the exterior could be represented in terms of Euclidean coordinates in some idealized space. This is precisely what is done when Euclidean coordinates are used to describe the surfaces of constant luninous density that obtain from photographs of galaxies – the warping of space by galactic matter is ignored.

The following parametric form of the equations of the isopleth, as follow directly from (31.3) and (29.2), are more convenient for actual calculation:

$$\begin{aligned} y^1 &= r \sin u^1 \sin u^2 \exp(\omega - \Psi), \\ y^2 &= r \sin u^1 \cos u^2 \exp(\omega - \Psi), \\ y^3 &= r(1-\varepsilon^2)^{\frac{1}{2}} \cos u^1 \exp(\omega - \Psi). \end{aligned} \tag{31.4}$$

Eqs. (31.4) *specify the isopleth in terms of the functions* ω *and* Ψ *in Euclidean three-dimensional space with Cartesian coordinates* $\{y^i\}$. We have thus obtained the geometric perturbations due to the Ψ function as well as that due to the ω function, the latter having been identified with the classical gravitational potential by (29.6).

Although we have used the space $\tilde{\mathscr{E}}$ in obtaining this result, it is easily seen that the validity of the italicized statement depends only on the properties of the subspace $\mathscr{S} \cap \tilde{\mathscr{E}} \cap \tilde{\mathscr{T}}$. Accordingly, since the metric structure of this subspace is identical with that obtained from (29.16), (30.1) and (29.3) for $\mathscr{S} \cap \mathscr{E} \cap \mathscr{T}$, the additional structures assumed for $\tilde{\mathscr{E}}$ and $\bar{\mathscr{E}}$ do not effect the final result. Further, the results of the external enveloping environment obtained in Chapter I strongly suggest the form of $\mathscr{E}$ that is obtained from (29.3), (30.1) and (29.16), particularly in that $\bar{\mathscr{E}}$ represents the gravitational field in first approximation and the conformal related structure given by (29.16). The implication is particularly strong in those instances in which we may take $B = 0$ in (29.25).

B. The Discretization Equation

All of our results thus far have been couched in terms of the function Ψ. If further progress is to be made, we must obtain explicit evaluations of this function, and this in turn requires that we solve the discretization equation.

32. Normalization

The discretization equation, that serves to determine Ψ, is given in the subsidiary space $\mathscr{C}$ by (27.9):

$$\bar{a}^{\Gamma\Sigma}(\bar{V}_\Gamma + \partial_\Gamma \omega - \partial_\Gamma \Psi)\partial_\Sigma \Psi + (C - 2\mu\Psi/\Theta)\exp(2\omega - 2\Psi) = 0 . \qquad (32.1)$$

When $\mathscr{C}$ is referred to a $*$-coordinate system, the $\bar{a}$'s are given by (28.6) and we have (see (28.3))

$$\partial_0 \Psi \overset{*}{=} \partial_0 \omega \overset{*}{=} 0 , \qquad \partial_\Gamma \mu = \partial_\Gamma C = \partial_\Gamma \Theta = 0 . \qquad (32.2)$$

For simplicity, we introduce the quantities $\{g_{\alpha\beta}\}$ be the relations

$$g_{11} \overset{*}{=} -\bar{a}_{11} , \qquad g_{12} \overset{*}{=} g_{21} \overset{*}{=} 0 , \qquad g_{22} \overset{*}{=} -\bar{a}_{22} . \qquad (32.3)$$

We then have, from the definition of the covariant derivative,

$$\begin{aligned} \bar{a}^{\Gamma\Sigma}(\bar{V}_\Gamma \partial_\Sigma \Psi + \partial_\Sigma \Psi \partial_\Gamma \omega) \overset{*}{=} & -g^{\alpha\beta}(\partial_\alpha \partial_\beta \Psi - \bar{\Gamma}^\gamma_{\alpha\beta} \partial_\gamma \Psi) \\ & - g^{\alpha\beta} \partial_\alpha \omega \partial_\beta \Psi \\ & - \bar{a}^{00} \bar{\Gamma}^\alpha_{00} \partial_\alpha \Psi ; \end{aligned} \qquad (32.4)$$

all other terms drop out by (32.2) and the form of the $\bar{a}$'s. From (28.6) and (32.3), it follows that

$$2\bar{\Gamma}^\alpha_{00} \overset{*}{=} g^{\alpha\beta} \partial_\beta \bar{a}_{00} \overset{*}{=} -2 g^{\alpha\beta} \exp(-2\omega) \partial_\beta \omega .$$

Substituting this result into (32.4) and noting that $\bar{a}^{00} \overset{*}{=} \exp(2\omega)$, we see that

$$\bar{a}^{\Gamma\Sigma}(\bar{V}_\Gamma \partial_\Sigma \Psi + \partial_\Sigma \Psi \partial_\Gamma \omega) \overset{*}{=} -g^{\alpha\beta}(\partial_\alpha \partial_\beta \Psi - \bar{\Gamma}^\gamma_{\alpha\beta} \partial_\gamma \Psi) ;$$

the $(\partial_\Gamma \omega)$-term drops out from (32.1)! We thus have

$$g^{\alpha\beta}(\partial_\alpha \partial_\beta \Psi - \bar{\Gamma}^\gamma_{\alpha\beta} \partial_\gamma \Psi - \partial_\alpha \Psi \partial_\beta \Psi) + (2\mu\Psi/\Theta - C)\exp(2\omega - 2\Psi) \overset{*}{=} 0 \qquad (32.5)$$

in $\mathscr{C} \cap \bar{\mathscr{T}}$ for the determination of Ψ.

The ensuing analysis is significantly simplified if we introduce a normalization by the relations

$$m_{\alpha\beta} \overset{*}{=} r^{-2} g_{\alpha\beta} . \qquad (32.6)$$

We thus have, by (28.6) and (32.3),

$$m_{11} \overset{*}{=} 1 - \varepsilon^2 \sin^2 u^1 , \qquad m_{12} \overset{*}{=} m_{21} \overset{*}{=} 0 , \qquad m_{22} \overset{*}{=} \sin^2 u^1 . \qquad (32.7)$$

The m's are thus the components of the coefficient tensor of the first fundamental form on the surface of the "unit oblate spheroid" S_ε in Euclidean three-dimensional space with eccentricity ε and semimajor axis equal to unity. Thus, on introducing the operators

$$\begin{aligned} \Delta_2(\Psi) &= m^{\alpha\beta} V_\alpha V_\beta \Psi = m^{\alpha\beta}(\partial_\alpha \partial_\beta \Psi - \overset{*}{\Gamma}{}^\gamma_{\alpha\beta} \partial_\gamma \Psi) , \\ \Delta_1(\Psi, \Phi) &= m^{\alpha\beta} \partial_\alpha \Psi \partial_\beta \Phi , \\ 2\overset{*}{\Gamma}{}^\gamma_{\alpha\beta} &= m^{\gamma\nu}(\partial_\alpha m_{\beta\nu} + \partial_\beta m_{\alpha\nu} - \partial_\nu m_{\alpha\beta}) , \end{aligned} \qquad (32.8)$$

on S_ε, the discretization equation is equivalent to the following equation in the two-dimensional space S_ε:

$$\Delta_2(\Psi) - \Delta_1(\Psi, \Psi) + r^2(2\mu\Psi/\Theta - C)\exp(2\omega - 2\Psi) \stackrel{*}{=} 0. \tag{32.9}$$

Since Θ, μ and C are constants, the substitution

$$\Psi = \varphi + \Theta C/2\mu \tag{32.10}$$

reduces (32.9) to the equivalent equation

$$\Delta_2(\varphi) - \Delta_1(\varphi, \varphi) + D\varphi \exp(2\omega - 2\varphi) \stackrel{*}{=} 0, \tag{32.11}$$

where the constant D is defined by

$$D = r^2\xi, \quad \xi = \frac{2\mu}{\Theta}\exp(-C\Theta/\mu). \tag{32.12}$$

The problem thus boils down to that of finding solutions of the *reduced discretization equation*, (32.11), that are single-valued and C^2 on the unit oblate spheroid S_ε.

33. Evaluation *D*

We first note that S_ε is a compact, simply connected and complete metric space without boundary. Bochner's theorem [8] thus holds, and hence any single-valued, C^2 function φ on S_ε satisfies

$$\int \Delta_2(\varphi) dS_\varepsilon = 0, \tag{33.1}$$

where the integration extends throughout S_ε and dS_ε denotes the differential surface element

$$\sin u^1 \sqrt{1 - \varepsilon^2 \sin^2 u^1}\, du^1 du^2$$

with respect to the $*$-coordinate system in use. When this result is used in conjunction with the reduced discretization Eq. (32.11), it is shown in Appendix D that there are three cases.

The first case is that for which $\varphi = 0$ and the value of D is arbitrary. The value of r is thus arbitrary, by (32.12), while (31.4) shows that the isopleth experiences perturbations solely as a consequence of the gravitational potential ω.

The second case is that for which $\varphi = \text{constant} \neq 0$ with $D = 0$. Eq. (32.12) shows that either the isopleth is of zero extent ($r = 0$), which we can immediately discount, or else we must have $\mu = 0$. Consequently, since μ has been identified with $-\bar{\rho}$, there is no jump in the mean mass density of a galaxy in this case. To the first approximation afforded by

[8] Bochner, S.: *Duke Math. J. 3,* 334 (1937).

the perfect fluid model, there is thus no jump discontinuity. It is therefore consistent to take $S_{\Gamma\Sigma} = Q_{\Gamma\Sigma} = 0$, $W_{\Gamma\Sigma} = 0$, and the physical situation corresponds exactly to the continuous perfect fluid model without boundaries of condensations. In the case of the external environment, the results $\bar{S}_{\Gamma\Sigma} = J\, a_{\Gamma\Sigma}$ shows that we have no jump discontinuities for the conformally homogeneous models. The first two cases thus correspond to most previously analyzed cases, such as the Schwarzschild interior and exterior solutions of a star and the continuous perfect fluid model.

The third case gives solutions that are nonconstant on S_ε and these obtain only if D is positive[9]. Since r^2 is positive, (32.12) gives $2\mu/\Theta > 0$ as the condition for nonconstant solutions. With the identification $\mu = -\bar{\rho}$, μ is negative, and hence we must have $\Theta < 0$.

In order to translate the requirement $2\mu/\Theta > 0$ into an interpretable form, we first combine (20.13) and (23.20) to obtain

$$\mu U_\Gamma U_\Sigma - \mu \Psi(a_{\Gamma\Sigma} - U_\Gamma U_\Sigma) + W_{\Gamma\Sigma} = \Theta\{R_{\Gamma\Sigma}(a) - \tfrac{1}{2}(R(a) + \Pi)a_{\Gamma\Sigma}\}\,. \tag{33.2}$$

When this equation is transvected with $\{a^{\Gamma\Sigma}\}$ and the trace condition (25.3) is used, we then have

$$\mu(1-2\Psi) + \Theta k = -\tfrac{1}{2}\Theta(R(a) + 3\Pi)\,, \tag{33.3}$$

and hence (25.8) and (32.10) give us

$$\frac{2\mu}{\Theta}(1-2\varphi) = -R(a) - \Pi\,. \tag{33.4}$$

Thomas[10] has shown that the Gauss-Codazzi equations remain valid when there are jump discontinuities in the second derivatives of the metric tensor but continuous first derivatives, and hence we have

$$R_{\Gamma\Sigma\Lambda\Xi}(a) = b_{\Sigma\Xi} b_{\Gamma\Lambda} - b_{\Gamma\Xi} b_{\Sigma\Lambda} - R_{ABCD} x^A_\Gamma x^B_\Sigma x^C_\Lambda x^D_\Xi\,.$$

Accordingly, since $R(a) = R^{\Gamma\Sigma}{}_{\Sigma\Gamma}$, we obtain the result

$$R(a) = b_{\Gamma\Sigma} b^{\Gamma\Sigma} - (b^\Gamma_\Gamma)^2 + R(h) + 2R_{AB} N^A N^B\,. \tag{33.5}$$

If the Einstein field equations are transvected with $\{N^A N^B\}$, we obtain

$$R(h) + 2R_{AB} N^A N^B = 2\kappa T_{AB} N^A N^B\,,$$

[9] See Appendix D for exact expressions for D in terms of ratios of integrals of φ and its derivatives.

[10] Thomas, T. Y.: *J. Math. Anal. Appl.* *7*, 225 (1963). We depend on Thomas' results here, in particular, Section 6, since Thomas has considered just the case in which the space is hyperbolic-normal and the hypersurface is timelike. Schouten's treatment assumes a positive definite first fundamental form, and hence requires continual modification if we are to make use of the results he reports.

and hence we have

$$R(a) = b_{\Gamma\Sigma} b^{\Gamma\Sigma} - (b^{\Gamma}_{\Gamma})^2 + 2\kappa T_{AB} N^A N^B \text{ [11]}. \tag{33.6}$$

A combination of (33.4) and (33.6) now gives us

$$\frac{2\mu}{\Theta}(1-2\varphi) = -\Pi - b_{\Gamma\Sigma} b^{\Gamma\Sigma} + (b^{\Gamma}_{\Gamma})^2 - 2\kappa T_{AB} N^A N^B . \tag{33.7}$$

The requirement $2\mu/\Theta > 0$ is thus equivalent to

$$0 < (b^{\Gamma}_{\Gamma})^2 - b_{\Gamma\Sigma} b^{\Gamma\Sigma} - \Pi - 2\kappa T_{AB} N^A N^B \tag{33.8}$$

for $\varphi < \frac{1}{2}$. In particular, the previous considerations show that φ *is other than a constant (and less than* $\frac{1}{2}$*) if and only if*

$$2\kappa T_{AB} N^A N^B < (b^{\Gamma}_{\Gamma})^2 - b_{\Gamma\Sigma} b^{\Gamma\Sigma} - \Pi . \tag{33.9}$$

Conservative estimates of the left-hand side of (33.9) from theoretical considerations (principally the perfect fluid model, or the plasma model) and the right-hand member of (33.9) from the known sizes and shapes of E and SO galaxies and the strengths of their gravitational potential[12] show that this condition is satisfied with reasonable values for the constant Π. We shall henceforth assume that the *nonconstancy condition* (33.9) is satisfied.

34. The Linearized Discretization Equation

An analysis of the reduced discretization Eq. (32.11), is given in Appendix D, where it is shows that there is a range of values of the L_2-norm of φ on S_ε about zero for which the linearized version of the reduced discretization equation yields correct values for both φ and D. We now confine our attention to the nonconstant solutions of (32.11) which are sufficiently "small" that the linearization is acceptable. By this we mean that

$$\int \varphi^2 dS_\varepsilon \ll 1 .$$

In this event, (32.11) is replaced by

$$\Delta_2(\varphi_0) + D_0 \varphi_0 \exp(2\omega) = 0 . \tag{34.1}$$

[11] Since $\{T_{AB} N^B\}$ is continuous across $\mathscr{S}$, as a consequence of the existence conditions (17.9), the evaluation of the quantity $T_{AB} N^A N^B$ is unambiguous even though $\mathscr{S}$ carries jump discontinuities in the momentum-energy tensor.

[12] The potential is required in order to estimate the timelike eigenvalue (associated with a timelike eigenvector) of the coefficient tensor of the second fundamental form; see, for instance, Thomas, T. Y.: *J. Math. Anal. Appl.* 7, Section 7 (1963).

Argument based on this approximation is, unfortunately, necessary, for the gross nonlinearity evidenced in the reduced discretization equation puts the problem of obtaining general solutions beyond reach at the present time. The reader is referred to Appendix D for the second and third order terms under expansion in the L_2-norm of φ.

In view of the previous evaluation of ω, it is seen that if the isopleth is not too near the nucleus of a galaxy, $\exp(2\omega) \approx 1$ to a high order of precision. We have, however, retained ω throughout the analysis, and hence let us set

$$\lambda^2 = D_0 \exp(2\bar{\omega}), \tag{34.2}$$

where $\bar{\omega}$ denotes the mean value of ω on $\mathscr{S}$. It then follows, from the known smallness of ω and from the even smaller variation of ω over the outer surface of E and SO galaxies, that

$$\Delta_2(\varphi_0) + \lambda^2 \varphi_0 = 0 \tag{34.3}$$

provides an excellent approximation of Eq. (34.1); that is, $\exp(2\omega - 2\bar{\omega}) \approx 1$. On combining (32.12) and (34.2), with D replaced by D_0, we finally obtain the following relation between the various parameters:

$$r = \lambda \sqrt{\frac{\Theta}{2\mu}} \exp\left(\frac{C\Theta}{2\mu} - \bar{\omega}\right). \tag{34.4}$$

Under the substitutions

$$\varphi_0 = \alpha G(z) L(u^2), \qquad z = \cos u^{1*}, \qquad \alpha = \text{constant}, \tag{34.5}$$

(32.7), (32.8) and (34.3) leads to

$$L'' + m^2 L = 0, \tag{34.6}$$

$$\begin{aligned} &\{1-\varepsilon^2(1-z^2)\}(1-z^2)G'' - z\{2-\varepsilon^2(1-z^2)\}G' \\ &\quad + \{1-\varepsilon^2(1-z^2)\}^2\{\lambda^2 - m^2(1-z^2)^{-1}\}G = 0. \end{aligned} \tag{34.7}$$

The above equations are to be solved subject to the conditions that the function $G(z)L(u^2)$ is nonconstant, single-valued, bounded and of class C^2 at all points on S_ε, where the coordinate ranges are

$$-1 \le z \le +1, \qquad 0 \le u^2 < 2\pi. \tag{34.8}$$

These conditions can be satisfied if and only if m, referred to as the *twig number*, is an integer and

$$\lambda = \rho(n,m,\varepsilon), \tag{34.9}$$

where the *branch number*, n, is an integer greater than or equal to $|m|$ and $\rho(n,m,\varepsilon)$ is the *discretization function*. This function is a continuous function of the argument ε over the domain $[0,1)$ for each admissible pair of values of m and n. With $m = \varepsilon = 0$, (34.7) reduces to the associated Legendre equation, and we have $\rho(n,0,0) = \sqrt{n(n+1)}$, while $\rho(n,0,\varepsilon) = \rho(n,\varepsilon)$ is given in Table 3 and Fig. 1 for $\rho(n,\varepsilon) \leq 9.0$[13]. Accordingly, (34.4) gives us

$$r = \rho(n,m,\varepsilon)\sqrt{\frac{\Theta}{2\mu}}\exp\left(\frac{C\Theta}{2\mu} - \bar{\omega}\right). \tag{34.10}$$

Table 3. $\rho(n,\varepsilon)$

ε \ n	1	2	3	4	5	6	7	8
0	1.4142	2.4495	3.4641	4.4721	5.4772	6.4807	7.4833	8.4853
0.05	1.4152	2.4511	3.4663	4.4750	5.4807	6.4848	7.4880	8.4906
0.10	1.4185	2.4559	3.4730	4.4835	5.4911	6.4971	7.5022	8.5067
0.15	1.4239	2.4641	3.4842	4.4979	5.5086	6.5178	7.5260	8.5337
0.20	1.4315	2.4756	3.5001	4.5182	5.5335	6.5471	7.5598	8.5720
0.25	1.4415	2.4907	3.5210	4.5449	5.5660	6.5855	7.6041	8.6221
0.30	1.4541	2.5095	3.5470	4.5783	5.6066	6.6335	7.6594	8.6848
0.35	1.4694	2.5324	3.5787	4.6188	5.6561	6.6919	7.7267	8.7610
0.40	1.4878	2.5596	3.6165	4.6672	5.7151	6.7615	7.8070	8.8520
0.45	1.5094	2.5917	3.6611	4.7242	5.7846	6.8436	7.9017	8.9592
0.50	1.5349	2.6291	3.7132	4.7910	5.8661	6.9398	8.0126	
0.55	1.5647	2.6726	3.7741	4.8688	5.9611	7.0519	8.1419	
0.60	1.5997	2.7231	3.8451	4.9596	6.0718	7.1827	8.2926	
0.65	1.6406	2.7818	3.9282	5.0656	6.2013	7.3354	8.4688	
0.70	1.6890	2.8504	4.0259	5.1901	6.3533	7.5149	8.6759	
0.75	1.7467	2.9311	4.1420	5.3375	6.5338	7.7278	8.9214	
0.80	1.8163	3.0273	4.2823	5.5145	6.7510	7.9838		
0.85	1.9012	3.1441	4.4558	5.7314	7.0184	8.2982		
0.90	2.0113	3.2909	4.6786	6.0058	7.3603	8.6977		
0.91	2.0370	3.3253	4.7315	6.0703	7.4413	8.7918		
0.92	2.0643	3.3618	4.7880	6.1389	7.5278	8.8919		
0.93	2.0934	3.4008	4.8487	6.2121	7.6205	8.9990		
0.94	2.1247	3.4427	4.9141	6.2907	7.7206			
0.95	2.1585	3.4881	4.9852	6.3758	7.8293		$\rho > 9.0$	
0.96	2.1952	3.5376	5.0631	6.4687	7.9487			
0.97	2.2357	3.5925	5.1496	6.5715	8.0813			
0.98	2.2811	3.6548	5.2475	6.6877	8.2319			
0.99	2,3338	3.7286	5.3626	6.8249	8.4095			
0.995	2.3649	3.7729	5.4311	6.9071	8.5155			

[13] The calculatory algorithm and numerical analysis leading to the explicit calculation of the discretization function and the associated solutions of (34.7) were developed by O. Gross, and is reported in the second Appendix of Edelen, D. G. B.: Possible Galactic Scale Discretization. The RAND Corporation, RM-3941-RC (1963).

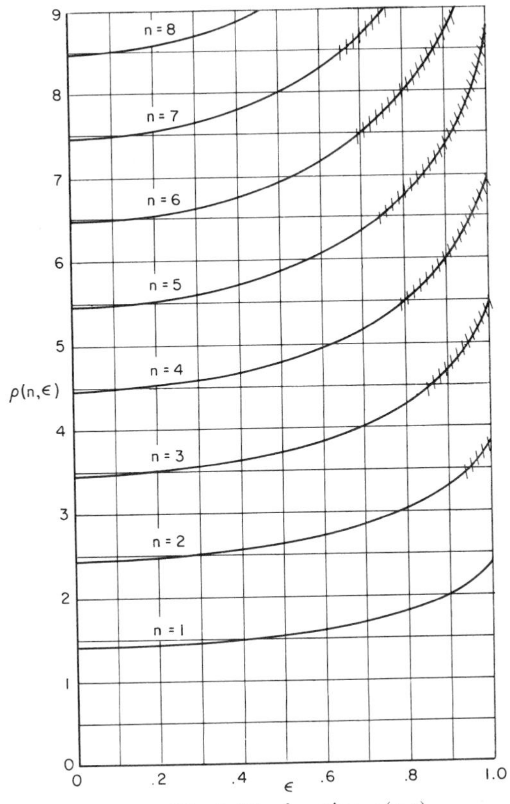

Fig. 1. The function $\rho(n,\varepsilon)$

The solutions of (34.6) need hardly be commented upon. With respect to (34.7), however, the story is another matter. For simplicity, we normalize the G-function by the requirement

$$G(1) = 1 \tag{34.11}$$

in the case where $m = 0$. It follows from (34.7) and (34.9) that $G(z)$ depends on the parameters n and ε, for $m = 0$, so we may write

$$G = G(z;n,\varepsilon)\,. \tag{34.12}$$

This function is unique. For given n and ε, under the condition that $G(z)$ shall be bounded for z in the closed interval $[-1,+1]$, $G(z)$ is an even or an odd function of z as n is an even or an odd positive integer. Figs. 2, 3 and 4 give $G(\sqrt{1-x^2};n,\varepsilon)$ for $n = 3$, 4 and 6, respectively. Here, x denotes the equatorial distance from the axis of rotational symmetry of the corresponding point on the unperturbed oblate spheroid S_ε $(z^2 = \cos^2 u^1 = 1-x^2)$.

C. Equilibrium Morphology of E and SO Galaxies and the Scale Question

We have obtained explicit equations for the isoplethic suriaces of galactic structures in states of geometric equilibrium and the corresponding linear scales as consequences of the conditions for the existence of solutions to the Einstein field equations in the presence of jump discontinuities in the momentum-energy tensor. This theory assumes the satisfaction of the isoplethic condition and the trace condition.

Almost all of the theoretical results obtain from the φ function, which must satisfy the reduced discretization equation. The solutions of this equation form three classes; the first two classes yield constant-valued φ functions and lead to physical situations similar to those treated in the literature. We accordingly confined our attention to those cases for which $\varphi \neq$ constant and for which φ had sufficiently small values that the linearized version of the reduced discretization equation was valid. The implications of the theory, as it pertains to the morphology and to the structure of the linear scales of E and SO galaxies, will now be obtained.

35. The Resulting Morphology of E and SO Galaxies

When Eqs. (31.4), (31.10), (34.5) and (34.10) are combined, the equations for the isopleths are given by

$$\begin{aligned} y^1 &= \rho(n,m,\varepsilon)\sqrt{\frac{\Theta}{2\mu}(1-z^2)}\,\sin u^2 \exp(\omega-\bar{\omega}-\alpha G(z)L(u^2))\,,\\ y^2 &= \rho(n,m,\varepsilon)\sqrt{\frac{\Theta}{2\mu}(1-z^2)}\,\cos u^2 \exp(\omega-\bar{\omega}-\alpha G(z)L(u^2))\,,\\ y^3 &= \rho(n,m,\varepsilon)\sqrt{\frac{\Theta}{2\mu}(1-\varepsilon^2)}\,z \exp(\omega-\bar{\omega}-\alpha G(z)L(u^2))\,, \end{aligned} \tag{35.1}$$

where

$$z = \cos u^1\,, \quad 0 \le u^1 \le \pi\,, \quad 0 \le u^2 < 2\pi \tag{35.2}$$

and $G(z)$ and $L(u^2)$ are solutions of (34.6) and (34.7) under the condition that $G(z)L(u^2)$ is a single-valued, bounded and C^2 function at all points on the unit oblate spheroid S_ε. We could discard the term $\exp(\omega-\bar{\omega})$ in the above equations since this term is unity to a high degree of precision on isopleths reasonably removed from the galactic nucleus. We have chosen to retain these terms, however, in the interest of completeness.

An examination of (34.6) and (34.7) shows that there are no single-valued functions of the form $G(z)L(u^2)$ on S_ε that are independent of the elliptic colatitude, u^1, other than the constant solutions[14]. Further, the elliptical and SO galaxies exhibit no perceptible deviations in their equatorial planes. We may thus confine our attention to the longitude-independent cases by setting $L(u^2) = 1$ and $m = 0$. It is for precisely this reason that we confined our attention in the previous Section almost exclusively to the case $m = 0$. We thus have the cases in which $\rho = \rho(n,\varepsilon)$ and $G(z)$ is given by solutions of (34.7) with $m = 0$. In the case where $\varepsilon = 0$, (34.7) reduces to the associated Legendre equation and we have $G(z;n,0) = P_n(z)$. The G-function can be calculated for nonzero values of ε and is depicted in Figs. 2, 3 and 4 for $n = 3$, 4 and 6, respectively.

The equilibrium isoplethic surfaces of E and SO galaxies are thus described by (35.1) with $\rho(n,m,\varepsilon)$ replaced by $\rho(n,\varepsilon)$ and $L(u^2) = 1$. Since

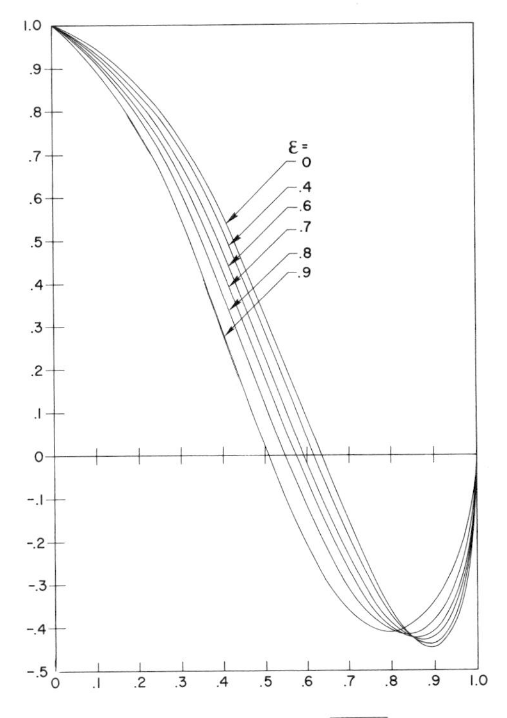

Fig. 2. The function $G(\sqrt{1-x^2};3,\varepsilon)$

[14] With $\varphi = \varphi(u^1,u^2)$, $\varphi(\pi,u^2)$ assumes all values between max $\varphi(\pi/2,u^2)$ and min $\varphi(\pi/2,u^2)$, while the parametric coordinates $(u^1,u^2) = (\pi,u^2)$ define the same point on S_ε for all u^2 in $[0,2\pi)$.

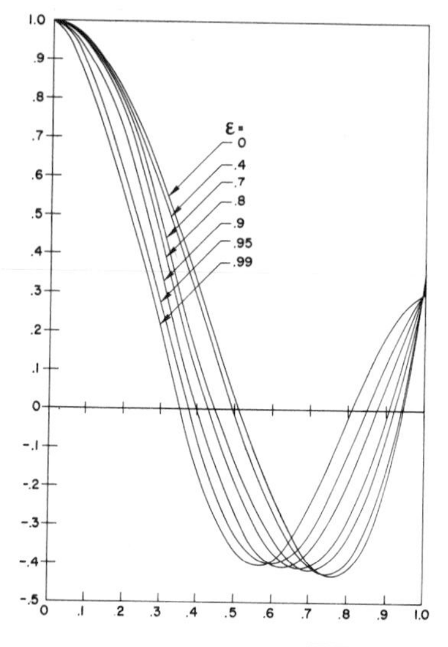

Fig. 3. The function $G(\sqrt{1-x^2};4,\varepsilon)$

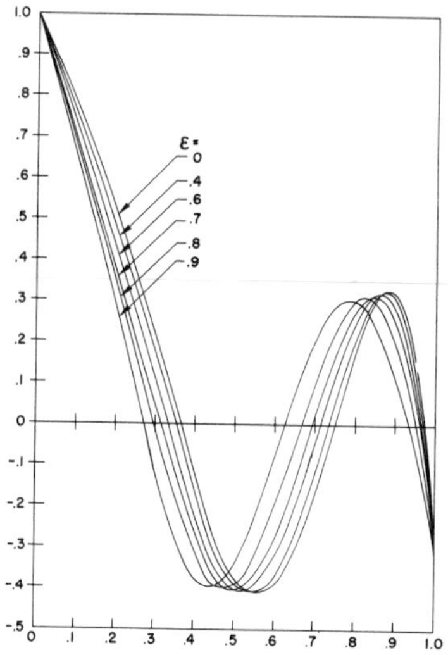

Fig. 4. The function $G(\sqrt{1-x^2};6,\varepsilon)$

no variation with longitude is considered, there is no lost of generality in confining the discussion to the $(y^1 = X, y^2 = Z)$-plane. Further, since the three-dimensional morphology depends only on the contours and not on the appropriate linear scale, we may plot the resulting contours on a single linear scale. Under the substitution, $\exp(\omega - \bar{\omega}) \approx 1$, the equilibrium isoplethic surfaces of E and SO galaxies are given by

$$\begin{aligned} X &= \sin u^1 \exp(-\alpha G(\cos u^1; n, \varepsilon)), \\ Z &= (1-\varepsilon^2)^{\frac{1}{2}} \cos u^1 \exp(-\alpha G(\cos u^1; n, \varepsilon)). \end{aligned} \tag{35.3}$$

The morphological classification parameters for the linearized, longitude-independent cases are therefore the two continous parameters α and ε and the discrete, integer-valued parameter n. In addition to these parameters, one must include the parameters that describe the effects of projection onto the plane of the sky, the inclination of the axis of symmetry, and the mass-luminosity ratio before direct comparison can be made with the observed isophotal contours of E and SO galaxies. We have not included the projection or mass-luminosity parameters [15], since their effects may be accounted for in any particular case once the inclination angle and the luminous density profiles are known.

[15] It is assumed that surfaces of equal luminous intensity can be identified with surfaces of equal mass density. This is necessary in view of the identification of the theoretical isopleths with surfaces of constant mass density by the relation $\mu = -\bar{\rho}$.

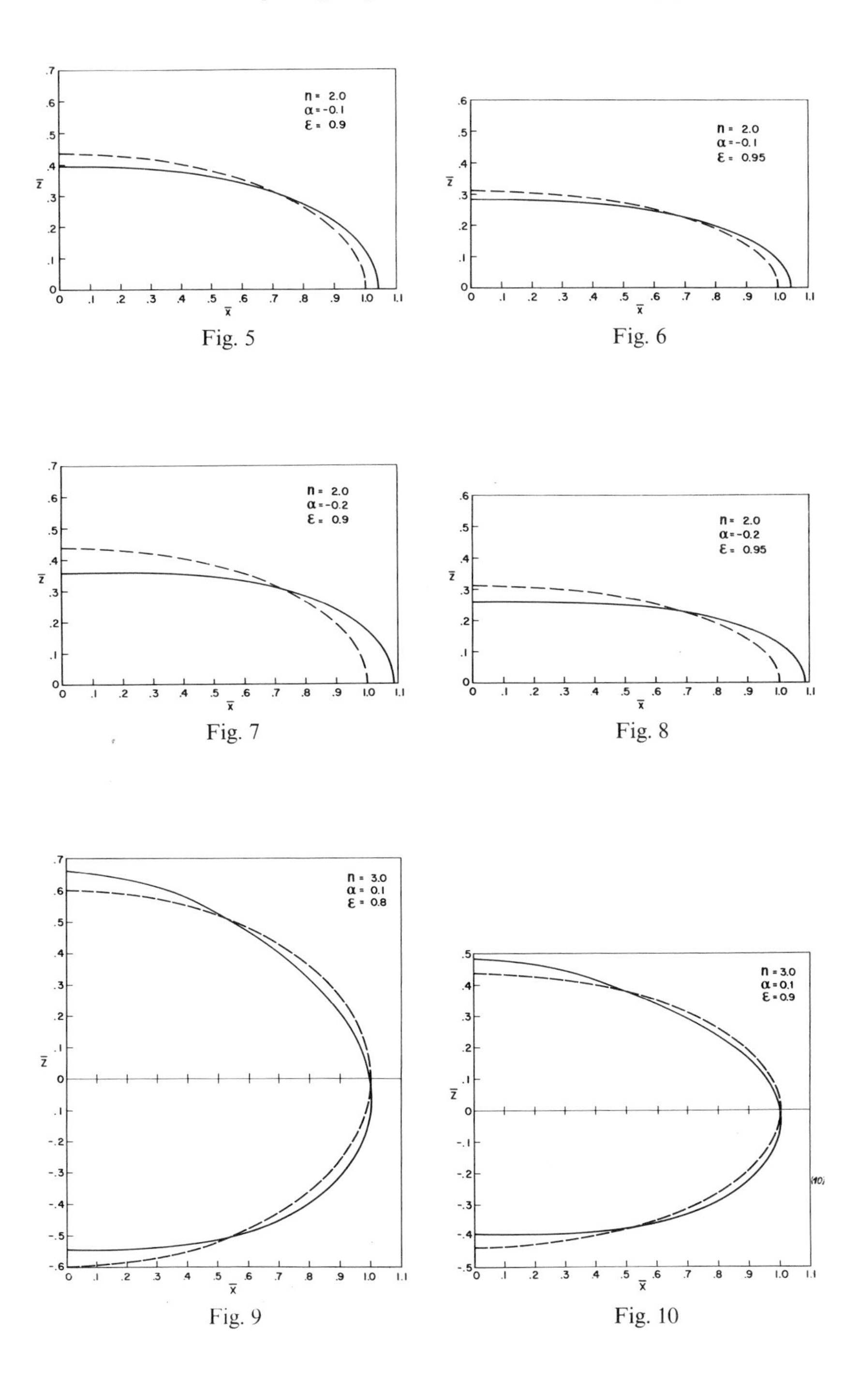

Fig. 5

Fig. 6

Fig. 7

Fig. 8

Fig. 9

Fig. 10

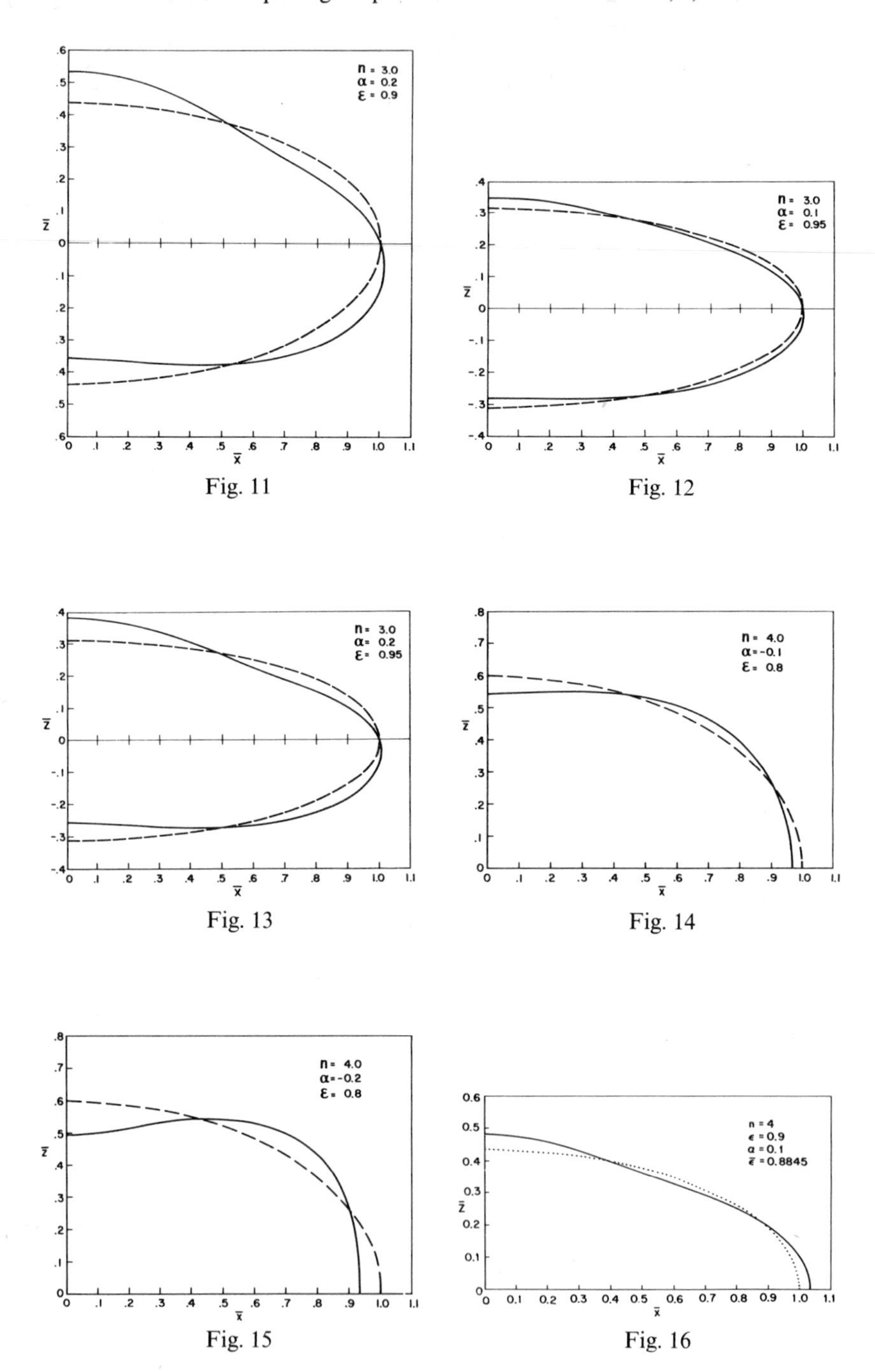

Fig. 11

Fig. 12

Fig. 13

Fig. 14

Fig. 15

Fig. 16

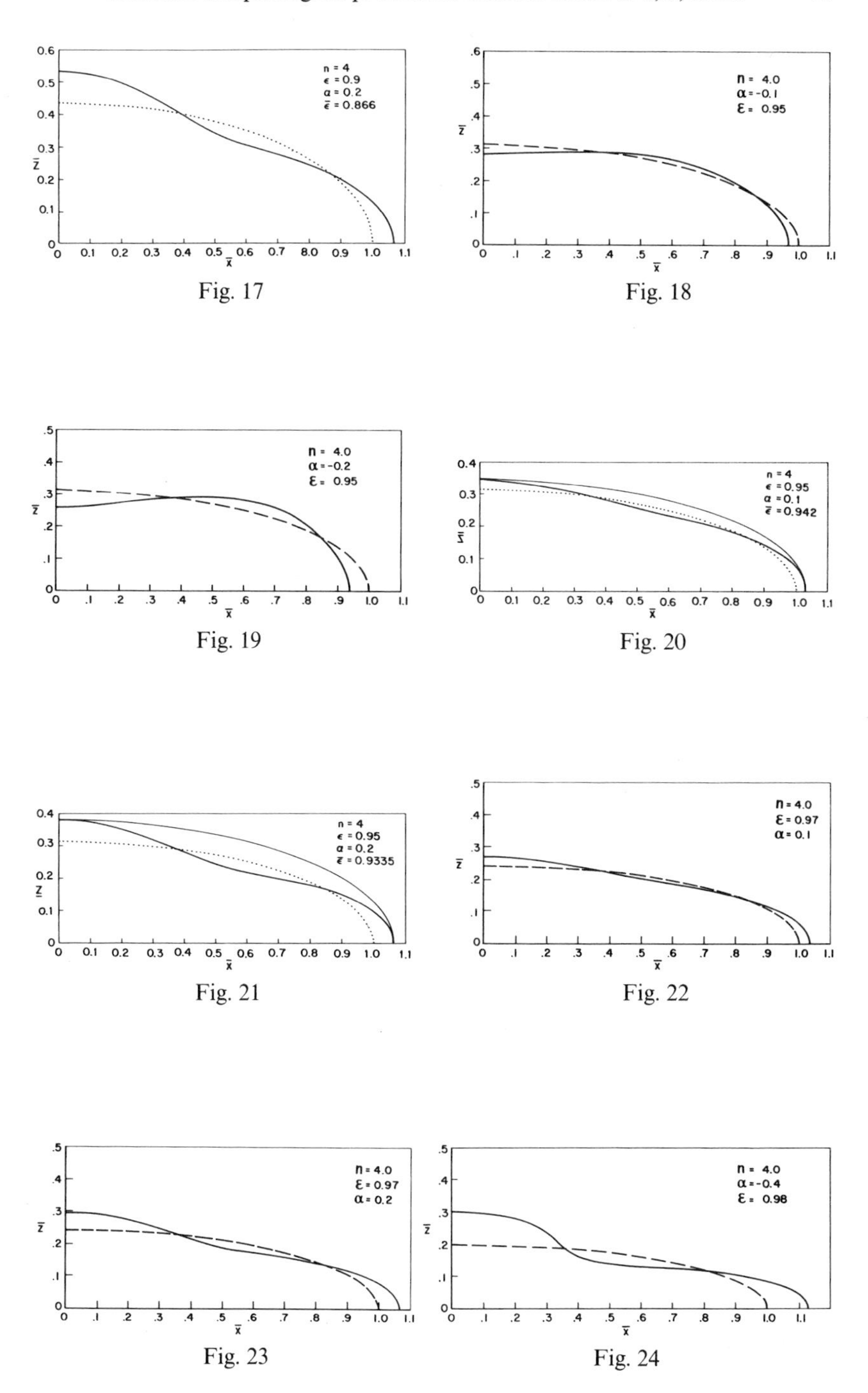

Fig. 17

Fig. 18

Fig. 19

Fig. 20

Fig. 21

Fig. 22

Fig. 23

Fig. 24

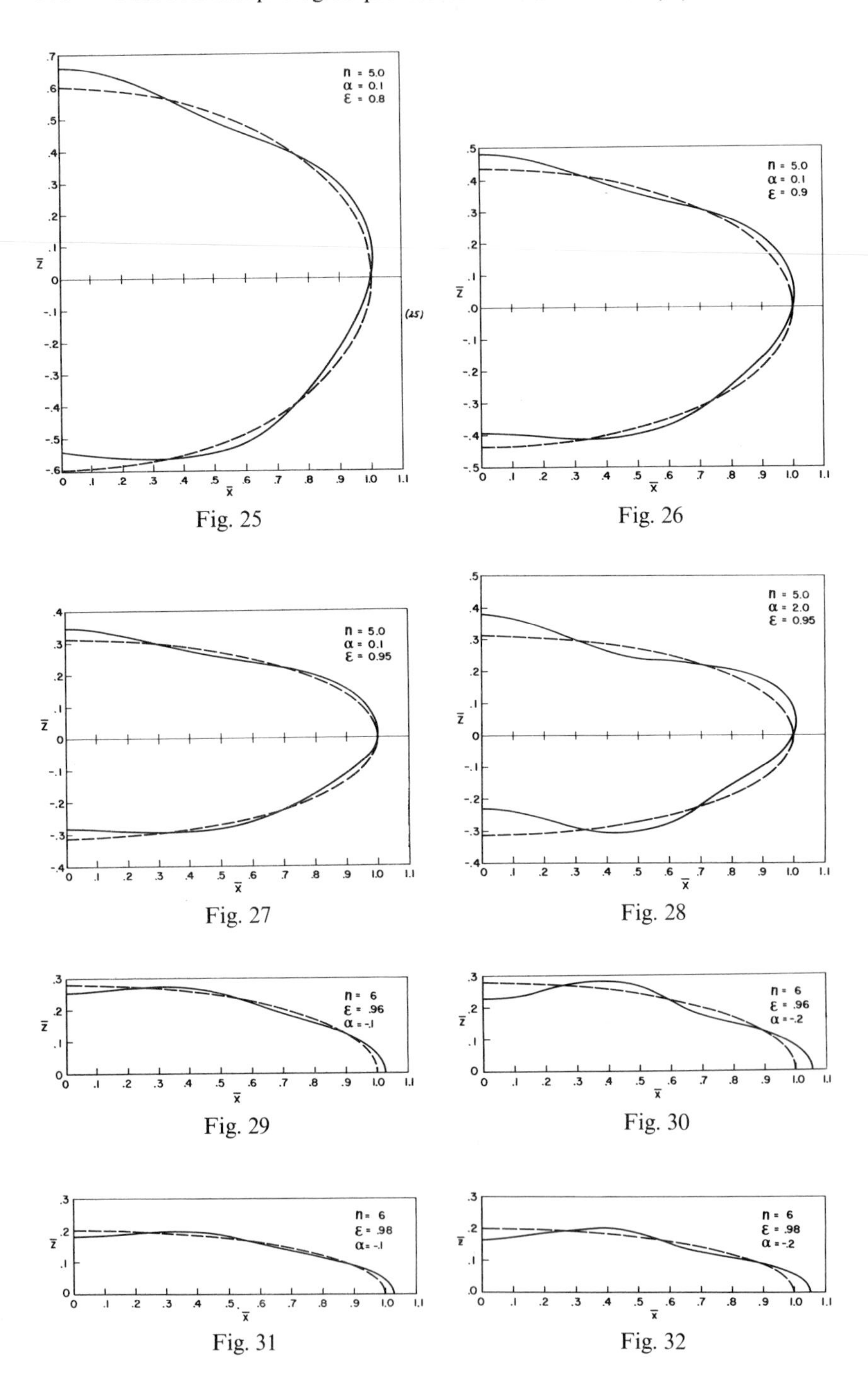

Fig. 25

Fig. 26

Fig. 27

Fig. 28

Fig. 29

Fig. 30

Fig. 31

Fig. 32

Plots of Z versus X in the first quadrant and in the half-plane $X \geq 0$ are given in Figs. 5 through 32 for various values of α, n and ε. The dashed or dotted curves represent the unperturbed ellipsoids that obtain for $\alpha = 0$.

It is hoped that in the future, isophotes of a large number of E and SO galaxies can be obtained and compared with the contours reported here[16]. We note at this point that the outlines of NGC 3115, NGC 128, NGC 7332 and IC 3973 are obtained for n equal to 4, 6, 6 and 3, respectively, with the appropriate values of α and ε[17]. For the classification schema that obtains under geometric equilibrium, the "peculiar box-shaped nuclei" of NGC 128 and NGC 7332 cease to be enigmatic, for they become equilibrium contours for $n = 6$, fit into an orderly sequence of shapes that progress out of the ellipticals, and have equilibrium interpretations as previously obtained only for the ellipticals. Further, for α sufficiently small, the classical outlines of the early and middle elliptical galaxies are obtained. As n increases, the value of α must decrease if the L_2-norm of φ on S_ε is to be held constant (constant error in the linearization of the reduced discretization equation). If thus follows, under the linearization assumption, that the larger ellipticals (larger n for constant values of the other parameters) should be true ellipticals to a higher order than the smaller ellipticals. We have, accordingly, confined our presentation of the predicted profiles to those for which $n < 7$. Several large values of α have been included, so that an indication of the nature of the α-dependence may be gleaned. These cases must be treated as indicative only, since they derive from a linearization that holds only for "small" values of α.

36. The Linear Scale

We now turn to the more controversial question of the linear scale that obtains for galaxies in states of geometric equilibrium. From (35.1) we see that the linear diameter, L, of an equilibrium isopleth is given by

---

[16] Several of the predicted equilibrium contours have been presented before Commission 28 of the I. A. U. at the 12th International Meeting in Hamburg in August of 1964. Bondi voiced the opinion, without having examined the theory, that the results must come from solving a wave equation with the wrong data, but an elementary examination of the results reported here show that this is not the case. There is indeed an equation involved of the wave type, but the condition that we have a time-oriented irrotational isometry gives the exact condition $\dot{\Psi} = 0$, and this implies that $\partial_0 \Psi \stackrel{*}{=} 0$; the wave operator reduces to Laplace's operator. A full listing of the predicted contours can be found in Edelen, D. G. B.: *J. Astrophysics Space Sci. 3*, 56 (1969).

[17] Compare the curves given here with plates 1, 4 and 6 of Sandage, A.: *The Hubble Atlas of Galaxies*. Washington: Carnegie Institute of Washington 1961.

$$L = 2\sqrt{\frac{\Theta}{2\mu}}\,\rho(n,\varepsilon)\exp(-\alpha G(0;n,\varepsilon))\exp(\omega-\bar{\omega}). \tag{36.1}$$

Accordingly, these diameters are the diameters of equilibrium E and SO galaxies under the present theory if the mass density ρ_0 used to define the isopleth (the luminous density I_0 used to define the isophote) is taken to be that value at which the measurements are taken. For EO galaxies ($\varepsilon = 0$, $E = 0$), the contribution from the exponentials is approximately one ($\alpha \approx 0$ for ellipticals and $\omega = \bar{\omega}$), and hence we have

$$L \approx \sqrt{\frac{2\Theta}{\mu}}\,n(n+1) \quad \text{for} \quad \varepsilon = 0. \tag{36.2}$$

We may thus conclude that *the diameters of equilibrium isopleths of E and SO galaxies are given by*

$$L = l\,\rho(n,\varepsilon), \tag{36.3}$$

where l is a scale factor that is determined by the jump, $-\mu$, in the mass density, the integration constant Θ, the quantity $\alpha G(0;n,\varepsilon)$, and the quantities ω and $\bar{\omega}$ according to the formula

$$l = \sqrt{\frac{2\Theta}{\mu}}\exp(-\alpha G(0;n,\varepsilon))\exp(\omega-\bar{\omega}). \tag{36.4}$$

That l is indeed a scale factor with the dimensions of a length follows from (33.4).

We have defined the physical isopleths of the theory as surfaces at which the mean mass density of galaxies achieves the value $\rho_1 = \rho_0 + \bar{\rho}$ above a background density ρ_0, and hence we have implemented the condition $\mu = -\bar{\rho}$. Thus, if two different isopleths of the same galaxy are considered, the isopleth with the larger value of $|\mu|$ (Θ and α assumed fixed and $\alpha G(0;n,\varepsilon)$ known) will have the smaller diameter. This is as it should be, for the larger value of $|\mu|$ corresponds to a larger value of the mean mass density and hence lies closer to the nucleus. Of more importance, however, (36.3) and (36.4) predict that, for given Θ, n, α and $G(0;n,\varepsilon)$, the diameter of an E or an SO galaxy will vary as the reciprocal of the square root of the mass density on the isopleth. This result shows that the region of validity of our analysis of the linearized case is that in which the mass density varies approximately as the radius to the minus two power. For E and SO galaxies, this region of validity comprises the outer regions where the galaxy merges with the background field. This is also as it should be, for if the jump discontinuity above the background value of ρ_0 were large, it is evident that the resultant model would supply a very poor representation of the actual state of affairs. On the other hand, with small jump discontinuities in the mass density, the model may be considered as a first step of a finite difference shell method of analysis of the galaxy under the Einstein theory, and as such leads to an acceptable result. Since our attention is confined to the outermost shell, the present theory may be looked upon as defining the consistent boundary conditions for a shell model analysis of E and SO galaxies; as such, all information specified on the outermost shell will propagate to some finite number of interior shells, and hence our results will apply to finite thickness shells of E and SO galaxies.

What has been done, in effect, is to obtain those surfaces $\mathcal{S} \cap \mathcal{T}$ in $\mathcal{E} \cap \mathcal{T}$ (in $\mathcal{R}(t)$) which, for given l, support bounded, longitude-independent, C^2 solutions of the reduced discretization equation; we have determined the *eigengeometries* of $\mathcal{S} \cap \mathcal{T}$ for which $\mathcal{S}$ is in a state of geometric equilibrium with respect to the Einstein field and for which the deviations from oblate spheroidal symmetry are sufficiently small that linearization is possible. The situation is in direct contrast to the eigenvalue problems usually encountered in mathematical physics, and is akin to that in which we are given the frequencies, density, and tension of a membrane and asked to find its shape under the assumption that its boundary is homotopic to a circle. It is also evident that the construction of a general solution to the reduced and linearized discretization equation by summing the various eigenfunctions (various solutions of (34.7)) with respect to n is *not* valid for eigengeometry problems. This follows immediately from the fact that each eigenfunction of $\Delta_2(\varphi)$ on S_ε is defined on a distinct surface in $\bar{\mathcal{E}}$ by (35.1) and hence a summation of such functions is meaningless since they do not have common domains in $\mathcal{E} \cap \mathcal{T}$. Thus, although eigengeometry problems employ methods similar in many respects to those employed in eigenvalue problems, the two kinds of problems are essentially distinct.

37. Galactic Scale Discretization

From the manner in which μ is defined in terms of the deviation from a background mean mass density, the value of μ can be fixed once and for all provided it is chosen small enough to place the isopleth on the outer fringes of the galaxy[18]. The values of the integration constants Θ and α may vary from galaxy to galaxy, however, and hence the scale factor l may vary from galaxy to galaxy[19]. Eqs. (36.3), (36.4) show that if

---

[18] It was shown in the previous Section that the region of validity of the linearized theory is where μ is small and the isopleth is on the outer boundary of the galaxy. The value to be chosen for μ, if there is to be comparison with theory and observation, will obviously depend on the observational definition of the outer boundary and on the resolution of the instruments of observation.

[19] Recall that the normalization $G(1; n, \varepsilon) = 1$ gives $G(0; n, \varepsilon)$ a fixed value for fixed values of n and ε.

Θ varies over a large range of values from one galaxy to another, so will the linear diameter, L. If this be the situation, no further information can be drawn from the theory and no further observational verification of discretization is possible.

As an interesting and reasonable working hypothesis, we consider the following circumstances: *There is a class C of galaxies that can be partitioned into a finite number of subclasses C_i; the mean value of the scale factor for galaxies of the ith subclass is l_i; and the dispersion about this mean value is small compared with*

$$l_i \min_{\varepsilon, n} \left[\rho(n+1,\varepsilon)-\rho(n,\varepsilon)\right].$$

For elliptical galaxies, the value of α must be sufficiently small that the contribution from $\exp(-\alpha G(0;n,\varepsilon))$ is negligible in (36.4). This follows from (35.1) if the y's are to be the coordinates of points on the oblate spheroid (the contribution from $\exp(\omega-\bar{\omega})$ being unity to a high degree of accuracy). *The diameters of elliptical galaxies of class C, that satisfy the conditions of geometric equilibrium, thus arise from discrete distributions with small superimposed dispersions.* It would be hoped that the various scale factors l_i could be correlated with observable differences among the ellipticals, such as colour, mass/luminosity ratio, etc.

For $E\,0$ galaxies of class C, the above hypothesis leads to the formula

$$L_i = l_i\sqrt{n(n+1)} \tag{37.1}$$

since $\rho(n,0)=\sqrt{n(n+1)}$. The march of L_i with ε can be obtained from

$$L_i = l_i\rho(n,\varepsilon) \tag{37.2}$$

and Table 3, and the values of l_i are related to the physical parameters by

$$l_i = \sqrt{\frac{2\Theta_i}{\mu_i}\exp(-\alpha_i G(0;n,\varepsilon))\exp(\omega_i-\bar{\omega}_i)}\,. \tag{37.3}$$

These formula are the basis for what we have termed *galactic scale discretization*[20].

We have obtained discretized linear diameters of galaxies and an associated discretized morphological classification of E and $S\,0$ galaxies through the occurence of n as one of the morphological parameters. It was also shown in Part A of Chapter I that the prestructure approximation for conformally homogeneous cosmological models gives a form of discretization through the eigen-sequence $j(j+2)$, and this eigen-

[20] The initial announcement of galactic scale discretization was made in August of 1963 at the meeting of the American Astronomical Society in Fairbanks, Alaska, see Edelen, D. G. B.: *Astron. J. 68*, 535 (1963), and Wilson, A. G.: *Astron. J. 68*, 547 (1963).

sequence occurs both in the scales of the inhomogeneities and in the decay times for each of the modes of the inhomogeneities. The underlying reason for these discretizations, from the mathematical point of view, is that the spatial differential operators that occur have closed domains for the independent (spatial) variables which are complete and without boundary. In the cosmological case, the space-like variables parameterized a three-space that was equivalent to the surface of the unit 4-sphere. For galaxies, the space-like variables parameterized a two dimensional metric space that was homeomorphic to the surface of the 3-sphere. The theoretical discretizations that we have exhibited in Part One are thus directly attributable to the topological closure of the spaces parameterized by the space-like independent variables. The closure properties of the space-like structures given here thus provide detailed, specific realizations of the ontological requirement of closure implied by the view of perception discussed in the Prologue, and bring forth relations between closure and discretization.

PART TWO

Observational Considerations

The second part of this volume is given to the exposition of observational researches into the nature of macroscale discretization phenomena. In the extension of scientific knowledge, observation and theory both operate in exploratory and supportive roles, but neither operates meaningfully for long without the other. In the exploratory mode, the task of observation is the detection of new entities, new patterns, and new relationships in experimental data. In the supportive role, the task of observation is to confirm or negate the patterns and structures predicted by theory. In the exploratory role, the task of theory is to extend its structure and predict a set of check points where observation may be summoned in support. In the supportive role, the task of theory is to organize and structure the findings of observation and unite them with existing theoretical structures. In the work described in this volume, theory and observation both play both of these roles.

In Chapter IV, several of the basic problems associated with discrete distributions and their testing are discussed. In Chapter V the data, the tests and the results of three sets of apparently discretely distributed macrophysical measurements are discussed. These include functions based on the angular diameters of bright galaxies, the angular diameters of cluster galaxies and the redshifts of galaxies. In the analyses of diameter data, the observations are supportive, in the analyses of redshift data they are exploratory, in the sense defined in the preceeding paragraph.

In addition to those cited in the general acknowledgements, the support of the following individuals to various parts of the observational program is gratefully acknowledged: M. L. Humason, W. C. Miller, and A. R. Sandage of the Mt. Wilson and Palomar Observatories and J. L. Carlstedt and G. Kocher of the RAND Corporation.

CHAPTER IV

Observational Problems of Macroscale Discretization

For inorganic as well as organic structures, the higher the level of organization, the larger the number of characterizing parameters and the greater the possible variety among individuals. Thus entities such as stars and galaxies, although characterizable by a relatively few parameters, none the less are more complex than their atomic constituents and exhibit a much greater range of difference between individuals than do their building blocks. In fact, except for special species of galaxies, such as EO's, the same uniqueness of individuals seems to obtain among galaxies as is evident among living organisms. Accordingly, macro-entities, one of whose characterizing parameters is limited to a set of discrete values, may not clearly exhibit discreteness because of the perturbations contributed by the remaining parameters. Only in the case of all other parameters assuming equal values, could there be good expectation of detecting a discrete distribution in the values of one of these parameters. In other words, whenever a description is multi-parametered, an observable that depends on more than one parameter is not likely to exhibit the effect of a single parameter with any degree of purity, and the type of pure discrete distributions such as are observed in atomic spectra are not expected unless all or most of the characterizing parameters are discretized.

In the case of galaxies, the EO's form a set in which many parameters are equal, and if discretization exists among cosmic entities, the EO galaxies would seem high priority candidates for its manifestation. On the other hand, the great variety exhibited by most other observable cosmic bodies argues against detection of discretization in any single parameter such as "diameter". The theoretical results of Section 37 are entirely consistent with these general remarks. Only in the limiting case of spherical symmetry do the equations assume a simple enough form to constitute a prediction of an observable discretization.

The approach to macroscale discretization must, therefore, be through the study of sub-classes in which as many parameters as possible may be assumed constant. If we are able to detect simple consistent discrete patterns in these sub-classes, we may be able to step-wise arrive at the combined effects of the parameters involved. In Chapter IV, Part A, some of the epistemological aspects of discrete distributions are amplified and in Part B, the design of tests by which discrete distributions can be detected and weighed are discussed.

A. Epistemological Viewpoints

The concept of discrete is used in two senses in this volume. First in contrast to continuous, and second in contrast to random. In Section 38, these distinctions are amplified and related to the epistemological viewpoints adopted in the Prologue.

Hypotheses are accepted or rejected on the basis of several jointly applied criteria. These include their internal consistency and their consistency with respect to other theories, their domain of empirical validity, the precision with which they predict or map phenomena, and their economy of representation. In Section 39, these ingredients of hypotheses are discussed as background to the design of tests.

38. The Discrete and The Continuous

The dichotomy, *discrete vis-a-vis continuous*, has to do with the mode of entitation. In an extreme conceptualization, discrete versus continuous is the distinction between "atomicity" or total separateness and independence of entity, and "fluidity" or connectedness and undifferentiability of entity. Since no entites are totally isolated, the question of discreteness has to do with perception. It depends on the choice of resolving power, the background cut-off level, the spatial and/or temporal amplitudes and periods of the fluctuation of some physical parameter and their acuities or rates of perceived change.

When acuities are large, observers tend to entitate in the discrete mode, when small in the continuous mode, and when acuities are moderate, there is a difference of opinion whether one is looking at a fluctuation in a distribution or at an entity. For example, if luminosity is taken as the fluctuating physical observable, stars are readily classified as entities, since the acuity or rate of change of luminosity at some point (the limb) is extremely steep. Galaxies still possess sufficient luminous acuity to be entitated in the discrete mode (and receive a name), although there are some who prefer to say galaxies would all merge if the background luminosity could be reduced. Clusters of galaxies, having still much lower luminosity acuities, began in the 1920's as fluctuations in the density counts of galaxies, by the 1940's they had been entitated in the discrete mode, not because the luminosity acuity had improved but because of the discovery of a correlation between luminosity (galaxy area density acuity) and dynamical parameters (galaxy motion acuity). In the 1960's, the acuities of higher order clusters as perceived by observers of human scale are not large enough to resolve the rather strong mode-of-entitation arguments now waging.

From this illustration, it is seen that the question of discrete vis-a-vis continuous is a question of interpretation of fluctuations. The smaller the acuity of the fluctuation of a parameter, the more difficult it is to incorporate the parameter into physical theory and it is only when there exist correlations with other fluctuations that a "phenomenon" emerges.

The supportive correlations may be in *content* (as in the case of the dynamical parameters of clusters mentioned above) or they may be in the *context*, i.e., in the patterns, equivalence classes, ordered sequences, etc. formed together with other entities. These contextual relations bring us to the second sense in which discretization is used in this book.

In addition to the meaning attached to *discrete* vis-a-vis *continuous* discussed above, by a discrete distribution is meant a distribution possessing sufficient regularity that it is readily distinguishable from randomness as for example, the *quantized* distributions of bound energy states in atoms.

The philosophical viewpoint this book accordingly assumes is that the question of the "ultimate nature of matter", be it "atomistic", reducible to some set of ultimate particles or continuous (the ultimate particles being but manifestations of fluctuations in a substratum) is improperly posed. The proper model depends on the manner in which the physical phenomenon is perceived, on the relative scales, resolving power and acuities of the observer and the observed. However, once an object has been entitated in the discrete mode, there remains the problem of the degree of regularity existing among the measurable quantities of the entities. The question of discretization is thus the question of the amount of regularity or level of structuring that is manifested or may be inferred from the observables.

39. Paradigmatic Inference

If we propose to investigate a question such as: "Are the values of the measured diameters, d, of galaxies given by a function, $d_n = d_0 f(n)$, that assumes only discrete values (where d_0 is a scale parameter with dimension *length* and $f(n)$ is a specified function of a variable that assumes only integral values)", we would like a "yes" or "no" answer. But even if this were purely a statistical question, we realize that the answer cannot be "yes" or "no", but must be given in terms of the degree of fit of d_n to a certain percent of the measurements and the probability that this degree of fit could be generated by a set of random measures with the same density distribution over the same range, etc. Any "yes" or "no" becomes a subjective reaction to the values of the statistical parameters.

But the question is more than a statistical question, it involves a model of physical phenomena and consequently raises concomitant questions concerning the internal consistency of the model, the implication of the model with respect to other physical constructs and vice versa. Thus, in addition to *precision* tests, there are *consistency* tests, and questions concerning the extent of the domain of consistency. Finally, there are *economy* tests, inquiring whether the number of inputs to the model – the number of assumptions and *ad hoc* postulates – are sufficiently small with respect to the output of the model – the number of measurements fitted, the number of phenomena explained, and most important, the number of predictions subsequently verified – to make the model one of utilitarian satisfaction. The final "yes" or "no" is a subjective choice dependent on all of these values.

The complexity of nature forces us to use constructs or models that are partial in the sense that their domain of validity is only over limited sets of observations. Despite our predisposition for monism, the entirety of experience must necessarily be spanned by the use of many models, frequently leaving sizeable interstices. In the overall growth of knowledge, whether through extension of a model or through building a new model, our habitual epistemological viewpoint requires that contiguity be preserved with the patterned growth outward from the existing main body of knowledge. Epistemological "islands" built around *ad hoc* hypotheses to explain apparently isolated sets of experience are not agreeable, whatever their economies or precision, because of their limited comprehensiveness or domain of applicability.

Both the observations and certain critical parts of the theory pertaining to discretization are *insular* in the sense that the observations are neither expected nor predicted by any theory except the one presented here. The small size of the available sample of observations contributes further to the insularity of discretization. In order to continue a meaningful investigation in the face of the limited domain of internal consistency, an approach "building out from the island" was taken. The inferences based on the internally consistent insular construct were derived and tested against observation, whether or not they were explicable in terms of the epistemological main body of astronomy. Such inferences, we term *paradigmatic inferences* from their primary concern with patterns and internal consistency.

One of the principle hazards of a limited domain of internal consistency is that a systematic error may be mistaken for a law of nature. (In fact, a pattern that obtains only over a limited sample we term a systematic error, while a pattern holding over a wide domain we call a law.) Accordingly, an insular body of knowledge may have as its sole subject the properties of certain systematic errors. This is however a respectable enough subject in science when it is recognized as such. Since "non-insular" extensions of theory derive great structural support from the consistency of the main body of knowledge to which they are attached, their comprehensiveness is assured and because of this, we frequently subjectively sustain them in spite of weakness in precision and their being fugitives from Occam's razor. Insular hypotheses on the other hand, to be taken seriously, must be precise, economical, and even elegant.

The result of these epistemological proclivities is that we tend to restrict the growth of knowledge to an epitactic process building upon and ever contiguous to a single, central mass. Experience that cannot be readily subsumed epitactically or by a theoretical tendril extending out from the main organism is usually ignored or denied. To prevent this, we require an epistemological approach that permits islands to be occupied before bridges from the mainland can be built. The material presented here, like all newly discovered results, is presented on the basis of paradigmatic inference. Any self consistent pattern of relations, capable of making testable predictions, acquires a satisfactory domain of internal consistency when the number of relations in the pattern reaches a certain richness. Thus the number of diverse paradigmatic linkages to observed events times the sample size, rather than the sample size alone becomes the measure of confidence in an insular model. If the criteria of precision and economy are met, then paradigmatic inference that leads to a sufficiently large domain of consistency, can create an acceptable insular model. Ultimate acceptance, however, depends on removal of the insularity.

B. Design of Observational Tests

A specific test for the discretization hypothesis defined by Eq. (37.1) is derived in Section 42. Prior to the formulation of this specific test, however, some basic considerations relating to definitions, precisions, and sample sizes that apply to all discretization tests are given in Section 40. Section 41 gives an overview of some of the statistical features of tests that are treated in detail in later sections.

40. Tests for Discretization Hypotheses

The observational investigation of macro-discretization phenomena may be conveniently considered on three levels associated with the following three questions:

1. What macrocosmic observables, if any, exhibit clumped or discretized distributions?

2. What discretized distributions, if any, manifest regularities in the locations of the clumps?

3. What observed discretized distributions, if any, are consistent with theoretically predicted distributions (e.g., those of Eq. (37.1))?

Different requirements for definition, precision of measurement, and sample size are involved in the testing of hypotheses based on each of these three questions. In the case of certain discretized parameters, as for example, diameters of galaxies, the requirement for a higher level of precision of measurement places increasing restrictions on the class of permitted operational definitions. In other words, in some operational definitions not only the *process* of measurement, but the *precision* of measurement enters the definition of the parameter being measured. (See Section 42.)

In general, the precision of measurement sufficient to test a hypothesis based on question one, is a precision allowing the establishment of a set of equivalence classes identifying sameness and non-sameness to within a tolerance that may be specified from statistical considerations of the degree of clumping and the separation of the clumps (e.g., the internal σ of clumps must be less than 4 times the separation of the mean). Going beyond this to a test of a hypothesis based on the second question, not only requires sufficient precision for establishing the reality of the clumps, but sufficient precision to make the observed mean values of the clumps sharp enough so that ratios of separations are not blurred. But in addition to a higher level of precision, the problem of detecting regularity in pattern requires a large enough balanced sample to exhibit a pattern that is not a picture of the effects of observational selection.

The *ab initio* detectability of a pattern is a function of its complexity which may be measured (a) by the number of sub-patterns present and (b) by the degree of coupling or dependence that holds among the sub-patterns. If, through some clue (such as a correlation), sub-patterns may be readily isolated, the algorithms for their construction may be simple enough to unravel, or the sub-patterns in turn may be isolatable into further sub-components. This process may continue until the prescriptions of economy of formulation impose an end, i.e., until the number of degrees of freedom employed exceeds a certain percent of the data points that are fit. The acceptance or rejection of a pattern thus depends on the simplicity of the algorithm by which a sufficiently large percent of the sample may be generated to within a prescribed tolerance. The exact relation between degree of complexity, size of sample, and precision of fit that makes for an acceptable construct is in the end a subjective matter[1]. All other things being equal, the decision for adoption may come to depend on the temporal order of the steps in the formulation.

Hypotheses to be tested relevant to question number three have an advantage over those of question two in the sense that a clue to the algorithm for constructing the pattern (or a sub-pattern) has been predicted by theoretical considerations. This has the effect of relaxing somewhat both the internal error or precision of measurement and the sample size requirements. But trade-offs to the easing of these requirements include the matter of identification of the symbols of the theory with operationally defined parameters. (See Prologue, Part C.) This can be done, only uncertainly, if tests of dimensionality and quantitative consistency are not applicable. Only through the overlap of a sufficiently large set of properties of the symbols and the observational parameters, such as identity of functional relation as pattern, can *candidate* identification be established.

Another property of discrete distributions is that they may manifest regularities in the values of a single parameter only, i.e., their structural relations are purely "auto-relations". Since science usually seeks to establish laws through correlations (or functional relations depending on the level of precision) between two or more observables, unless the auto-regularities are correlated with some second parameter, their value is scientifically limited. Examples of auto-relations such as the Titius-Bode Law have not proven scientifically useful, because they provide no direct entry into a "physical" model. Thus while the statistical validity of auto-relations in physical parameters may be as substantial as that of multi-parameter relations, their interpretation is necessarily more ambiguous.

41. Remarks on Statistical Tests

There is a wide range of opinion governing the proper use of statistical tests in scientific inference[2]. In the case of models governing discrete distributions, the proper way to assess rival hypotheses through statistical analysis of data is an unsettled matter needing additional investigation. There are not even any standard tests available for assigning a significance to the degree of fit to a *discretized* null function.

---

[1] We cannot presume that comprehensiveness and precision are independent just as field of view and resolving power are jointly bounded by a limitation on the information capacity of the channel. Also in a general sense, Gödel's Theorem implies a trade-off between completeness and consistency. Such tradeoffs preclude monism in epistemology except as may be achievable through a multi-level approach. (See Bunge, M.: The Metaphysics, Epistemology and Methodology of Levels. In: *Hierarchical Structures*. Eds. Whyte, L. L., Wilson, A., and Wilson, D. New York: American Elsevier 1969).

[2] Edwards, A. W. F.: *Nature 222,* 1233 (1969).

The procedure followed in Sections 44 and 46 to compare the degree of fit between data and theory with the degree of fit between samples of random numbers and theory is a straight forward Monte Carlo method, which may be described briefly as follows: Let there be a set of N observations, such as diameters, designated by $S_1 \dots S_n$ distributed over a range $R = S_{max} - S_{min}$. Let there be M theoretical values, such as those proportional to $\sqrt{n(n+1)}$, designated by $t_1 \dots t_m$ included in the same range R. A fit of order (p_0, q_0) in terms of two parameters will be defined over R as the occurrence of p_0 members of the set $\{S\}$ differing from q_0 members of the set $\{t\}$ by less than a specified δ. A set $\{r\}$ of N random numbers distributed over the range R according to some specified density function may then replace the set $\{S\}$. A fit of order (p_r, q_r) between the random set and the set $\{t\}$ will be defined in an identical manner. The relative orders of fit (p_0, q_0) and (p_r, q_r) may be used as a measure of the chance fit of the data to the theory.

The order of fit (p, q) will be a function of N, the sample size; M/R, the mean density of theoretical points; and δ the specified bound on the residuals. With the sample size and range specified, how should M's and δ's be selected? It is evident that for any given δ, by making M large enough, the order of fit may be increased to include the whole sample and, for a given M by taking δ large, the order of fit may be increased to the whole sample. But beyond certain bounds these orders of fit are misleading. In general, to select δ either smaller than the uncertainty, e, in the observations or larger than the mean separation R/M is apt to lead to erroneous conclusions.

A feature of discretized distributions of additional help in evaluating significances of fits, are the *voids* between the points. The order of fits of the voids may be treated similarly to the order of fits of the observations.

A direct test of discreteness proposed by Efron[3] is based on the differences $y_i = D_{i+1} - D_i$ in the ordered sequence $n(D)$, from smallest D_1 to largest D_N. To a first approximation these differences will be exponentially distributed with mean and variance equal to $1/N f_i$ where f_i is the underlying distribution (constant, if $n(D)$ is uniform). If $N > 100$, the correlation between y_i and y_j is small, and the chi-square test for uniform distribution over the range 0 to 1 is

$$\sum_i (\log y_i - \log N)^2 = \chi_N^2(1.62).$$

A different test based on the *amount of clumpiness* has been proposed by Harris[4] for the analysis of discrete distributions that provides a method for hypothesis ranking.

[3] Page, T.: Proceedings of the Conference on Discrete Parameters in Cosmology. RAND Corp Memorandum RM-4267-RC, p. 12 (1964).

[4] *ibid.*

It must be remembered, however, that statistical inference like inductive inference, is asymmetric. While validity of a hypothesis in a large number of cases does not inductively prove a hypothesis, but only raises the level of confidence in it; invalidity in *one* case may disprove the hypothesis. Similarly, statistical inference cannot be used by itself to establish the validity of a hypothesis but it is very useful in throwing out hypotheses that are invalid.

The difficulties involving probability arguments were lucidly described by Neyman [5] in his account of the controversy between Joseph Bertrand and Emile Borel. "Whatever the observed configuration of objects, it is always possible to invent a characteristic of this configuration such that the probability of obtaining it by chance will be as small as desired. In other words, among points distributed by chance on a plane one inevitably obtains configurations some features of which are extremely improbable. Thus, probability of a particular observed feature does not constitute convincing ground for denying the chance mechanism underlying the distribution of points. Bertrand's pessimistic conclusion was that no probability argument could be useful in testing hypotheses. Borel disagreed; according to his intuition, probability theory could be used for testing hypotheses provided that certain precautions are observed. Two of these precautions are (1) that the test criterion be chosen *before* observations are started and (2) that the test criterion be a function of the observable variables "en quelque sorte remarquable"."

It is in establishing the meaning of "en quelque sorte remarquable" that the difficulties lie.

[5] Neyman, J.: *Astron. J. 66,* 557 (1961).

CHAPTER V

Observational Evidence for Discretization Phenomena

Two sets of observables are involved in the investigations of the macroscale discrete distributions described in this chapter. These are the angular diameters of galaxies and their redshifts. It has long been assumed that these parameters are independent, but evidence is given in the case of cluster galaxies for the existence of their interdependence. This interdependence leads from the discretization of diameters, discussed in Part A to the discretization of cluster redshifts, taken up in Part B.

The recent recognition of discretization effects in the redshifts of quasars and radio sources has greatly expanded the limited sample of observables available for testing. However, the cluster redshift discretization together with the quasar redshift discretization and their relation to one another require extensions of the basic discretization theory.

The fact that the two basic observables used in discretization studies are dimensionless, gives rise to the possibility that the fundamental dimensionless discretization scale parameters might be related to other basic dimensionless constants. Some results bearing on this matter are introduced in the concluding Section.

A. Diameters of Galaxies

The original detection of the possibility of macrocosmic discretization was in the diameters of EO galaxies. Because of the small size of the available sample, little progress was made with the investigation of discretization until the theoretical results of Section 37 became available, relating the relative sizes of the various diameter classes. The investigations then turned toward testing the diameter data for distribution functions based on sequences of the form $\sqrt{n(n+1)}$, $n = 1,2,3,\ldots$

Section 42 discusses problems implicit in definition and measurement of galaxy diameters; Section 43 tests a sample of bright nearby EO galaxies for discretized distributions; and Section 44 investigates discretization among galaxies in eight rich clusters.

42. Measurement of Diameters

The problems incumbent in the definition and measurement of elliptical galaxies are well known and extensively discussed in the literature[1]. If galaxies were balls or disks with sharply defined edges, there would be less difficulty. But galaxies are stellar aggregates whose density and luminosity vary throughout. In general, the distribution of luminosity is such that there is a bright core or nucleus surrounded by a region in which the brightness gradually fades out. But neither the core nor the outer portion is sharply defined and the real spatial extent of a galaxy is unknown. In fact, what is meant by spatial extent is a paraphrase of the problem. A *physical* definition for diameter of a galaxy might be: The distance from the center to the isopleth where the density drops to a specified value such as 1/2, 1/e, or 1 per cent of the central density. For diffuse objects with monotonically outward decreasing densities, definitions of diameter must take some such form. However, we don't measure the densities of galaxies themselves but rather we measure observables such as images on photographic plates or photocell tracings. Several ways of defining diameters using such observables have been proposed[2]. These definitions are operational definitions in the sense that they involve a prescribed process of measurement on the direct photographic image, on isophotal contours, or on luminosity profiles, etc. Because of the nature of the images, many difficulties are involved in obtaining measurements that are accurate reflections of the relative apparent sizes of galaxies. Systematic errors arise from seeing, exposure time effects, sky brightness, etc. In addition, elliptical galaxies come in several forms. These are not only the morphological differences of size, mass, and ellipticity, but other differences, such as degree of compactness or diffuseness and higher order departures of isoplethic surfaces from ellipsoidal symmetry. Thus, the fundamental concept of size becomes difficult to abstract in a uniform manner.

Whereas operational diameters such as those defined in terms of isophotes are suitable for comparison of galaxies located at a single distance (as is approximately the case for galaxies in the same cluster), such angular diameters do not necessarily possess the essential metric property of being inversely proportional to distance. In order to compare diameters of galaxies at different distances, r, the observed angular diameters (however operationally defined) must be readily reducible to a *metric* angular diameter, $\theta(r)$, with the property

[1] See for example, Hubble, E.: *Astrophys. J. 71,* 231 (1930); Holmberg, E.: *Lund Medd.* II, No. 128 (1950); Vaucouleurs, G. de: In: *Handbuch der Physik*, vol. LIII. Berlin-Göttingen-Heidelberg: Springer 1959.

[2] *ibid.*

$$\theta(r_1)/\theta(r_2) = r_2 f(z_2)/r_1 f(z_1)$$

where $f(z)$ is some function of the redshift, z, that depends on the properties of cosmic space. In other words, diameters useful for testing the existence of theoretical size distributions must functionally depend on distance through the relation, $[rf(z)]^{-1}$. The possibility of evolutionary changes in galaxy size and luminosity further contribute to the uncertainties in the proper choice of $f(z)$. All of these difficulties have resulted in diameters of galaxies being of limited use in cosmological research [3].

In consequence, the three steps to be taken in order to test a prescribed diameter discretization hypotheses are: First, the establishment of a well defined operation by which may be obtained a consistent set of measurements to be used as *candidate* diameters. Second, the reduction of this set to a set of *metric* angular diameters through corrections for systematic errors [4]. Third, the determination of the distance functions $r(z)$ and $f(z)$ so that distance effects can be removed and the angular diameters reduced to linear diameters. The reduction to linear diameters will be developed in more detail at the end of this section. Finally, the testing of the theoretical discretization function through comparison of the distribution of the set of linear diameters with the theoretically prescribed distribution is straight forward in the event of a one parameter distribution such as the $[n(n+1)]^{\frac{1}{2}}$ factor in Eq. (37.1). However, in the event of a two parameter distribution as would occur through a plurality of l_i classes in Eq. (37.1), the comparison becomes more difficult and special comparison procedures must be designed. It should be noted that if diameters are discretized and if it is possible to identify correctly observational and theoretical discretization classes, systematic errors in operational diameters may much more easily be isolated and diameters calibrated making use of the resolved size and distance effects.

In addition to the effects so far discussed, it will be necessary to consider the effects of unknown orientations of galaxies on the selectability of a suitable sample of ellipticals for testing Eq. (37.1) which applies to EO galaxies and, to within determinable deviations, to ellipticals of small true eccentricity. Samples of elliptical galaxies to be tested must be selected on the basis of their *apparent* eccentricities. Hubble has shown [5] that in any sample of elliptical galaxies with random orientation of axes of symmetry that only 55% of the apparent EO's are true EO's and the remainder are ellipticals with larger true eccentricities so oriented as to give an apparent *ellipticity* from 0 to 0.05

[3] Hubble, E., Tolman, R. C.: *Astrophys. J. 82,* 302 (1935); North, J. D.: *The Measure of the Universe.* Oxford: Claredon Press 1965.

[4] Vaucouleurs, G. de: *Ann. Astrophys. 11,* 247 (1948).

[5] Hubble, E.: *Astrophys. J. 64,* 321 (1926).

(corresponding to a range in eccentricity from 0 to 0.3)[6]. It is thus probable that any sample of low apparent eccentricity galaxies will be "contaminated" with galaxies of larger true eccentricity for which Eq. (37.1) does not hold. It is therefore necessary to investigate the probable degrees of contamination in our samples in order to determine whether or not discretization tests might be vitiated by this effect.

First, let us specify more precisely the degree of fit of Eq. (37.1) in terms of ε, the *true* eccentricity. The deviations in diameter from the values given by (37.1), as functions of eccentricity, ε, may be computed from the theoretical eigen values, $\rho(n,\varepsilon)$ assumed to hold for elliptical galaxies. On the basis of the set of theoretical eigen values given in Fig. 1, it can be shown that the percent increase in the major diameter of a galaxy over the $\varepsilon = 0$ diameter given by (37.1) for all values of n is less than 1% for all $\varepsilon \leqq 0.2$ and less than 5% for all $\varepsilon \leqq 0.42$. Thus, the testable sample for Eq. (37.1) consists not just of true EO's but of all ellipticals whose true eccentricity is less than or equal to a value corresponding to some specified deviation, usually selected to be the same as the internal error of measurement. For example, if the diameters are known to within 5% then Eq. (37.1) may be assumed to hold to the same precision out to true eccentricities of 0.42, etc.

With the test sample defined in terms of the true eccentricity and degree of precision, it is now required to define the sample in terms of the *apparent* eccentricity. The apparent eccentricity, ε_a, and true eccentricity, ε, are related by

$$\varepsilon_a = \varepsilon \sin \varphi$$

where φ is the angle between the rotational axis of the galaxy and the line of sight. Let us assume, for illustrative purposes, that the acceptable precision is 5%. Then any galaxy with an apparent eccentricity of less than 5% will be admittable to the test sample. In order for ε_a to be $\leqq 0.05$,

$$\sin \varphi \quad \text{must be} \quad \leqq 0.05/\varepsilon\,.$$

For $\varepsilon_a \leqq 0.05$, the values of $\sin \varphi$ and φ corresponding to different ε's are given in Table 4. The dotted band dividing Table 4 corresponds to $\varepsilon = 0.42$ which is the limiting value of ε for a galaxy to be testable to within a precision of 5%. All galaxies above the dotted band are testable regardless of orientation. The galaxies below the dotted band will not be testable since their major axes deviate by more than 5% from the values of Eq. (37.1). Those below the dotted band and whose φ's are greater than those given in the right column of Table 4 will have values

[6] If a is the major diameter and b the polar diameter of an elliptical galaxy (all assumed to be closely oblate spheriods), then the *ellipticity* is defined to be $E = 10[(a-b)/a]$ and the eccentricity to be $\varepsilon = (a^2-b^2)^{1/2}/a$.

of $\varepsilon_a \geqq 0.05$ and hence will be detected and rejected from the sample. The remaining galaxies, those below the dotted line with φ's less than given in the right hand column of Table 4, will neither be testable nor detectable as interlopers and will constitute the contaminating portion. The contaminating set is thus a subset of the set of galaxies with $\varepsilon > 0.42$ for which $\varphi \leqq 6°.8$.

Table 4. *Ranges of* $\sin\varphi$ *and* φ *for* $\varepsilon_a \leqq 0.05$

ε	$\sin\varphi$	φ
0.0	0 to 1.000	0 to 90°
0.1	0 to 0.500	0 to 30°
0.2	0 to 0.250	0 to 14°.5
0.3	0 to 0.167	0 to 9°.6
0.4	0 to 0.125	0 to 7°.2
0.42	0 to 0.119	0 to 6°.8
0.5	0 to 0.100	0 to 5°.7
0.6	0 to 0.083	0 to 4°.8
0.7	0 to 0.071	0 to 4°.1
0.8	0 to 0.063	0 to 3°.6
0.9	0 to 0.056	0 to 3°.2

If we assume that the axes of symmetry of galaxies are randomly oriented, then the probability of an axis falling in this contamination cone will be

$$1-\cos\varphi\,.$$

Thus for a precision of 5%, the percentage of undetectable contaminating galaxies will be

$$1-\cos 6°.8 = 0.7\%\,.$$

The percent contamination for various precisions is given in Table 5. The small percentages of contamination as shown in Table 5 which are based on random orientation of axes allow the discretization hypothesis to be restated: Eq. (37.1) is valid for *apparent* $E0$'s and low *apparent* eccentricity samples of ellipticals with exceptions determinable from the specified precisions given in Table 5.

Let us now return to the problem of the correction for distance effects. The conversion of corrected angular diameters to linear diameters involves those cosmological questions which arise in the comparison of objects located at different distances whose observed radiation originated at different times. Since it is desirable to hold the number of additional assumptions entering into the test hypothesis to a minimum, it will be best initially to test samples of galaxies for which cosmological effects are negligible. These samples will in general be either galaxies with small redshifts or galaxies located in the same volume of space, such as members of the same cluster. If then it develops that discretization effects do exist and are readily observed in these samples, it is evident that we will have at our disposal a yardstick for calibration of linear sizes of galaxies and hence possess a tool for cosmological exploration when samples at different distances are compared.

Table 5. *Percent contamination*

Precision (%)	φ (Degrees)	Contamination (%)
1	3.0	0.1
2	4.25	0.3
3	5.75	0.5
4	6.0	0.6
5	6.8	0.7
10	10.0	1.5

First, let us consider the samples of galaxies which should be relatively free of cosmological effects. The first sample consists of large bright nearby galaxies. The problem of distribution of these galaxies in distance can be met by assuming that the redshift is closely a linear indicator of distance and that the peculiar component of the redshift due to dynamical effects other than the general expansion is negligible. This evidently is not a bad assumption. The size of the peculiar component has been estimated by de Vaucouleurs to have a maximum value of 60 km/sec and by Humason, Mayall, and Sandage for most of their redshifts to have a value of 35 km/sec [7]. If these bounds are correct, then with redshifts greater than 1200 km/sec, the errors from the peculiar component should be less than 5%. However, caution should be exercised in assuming the peculiar component to be negligible in cases of closely interacting objects where large dynamical velocities may be present. There is also the possibility of a component in the observed redshift attributable to a suspected rotation of the local supercluster of galaxies for which a good correction is not yet available.

Under these assumptions, the conversion of angular diameters to linear diameters is straight forward. Assuming the Hubble velocity-distance relation in the form

$$H\Delta = cz$$

[7] Vaucouleurs, G. de: *Astron. J. 63,* 253 (1958); Humason, M. L., Mayall, U., Sandage, A. R.: *Astron. J. 61,* 97 (1956).

where H is Hubble's parameter, Δ is the distance, c the velocity of light and $z = \delta\lambda/\lambda$ is the spectral shift; the linear diameter S should be given to good approximation by

$$\log S = \log\theta + \log cz - \log H$$

where θ is the angular diameter. The discretization properties of Eq. (37.1) may be tested against S's for $E\,O$ galaxies or other suitable members of the sample for which θ and z are known. However, diameters of galaxies should be measured as reduced to a given cosmic time and not as they appear to a particular observer. Hence, in order to eliminate preferences given to the observer's particular location in space, it is necessary to correct the observed quantities for light travel time. The Δ, S, and θ of the above equations accordingly must represent the distance, the linear diameters, and the angular diameters reduced to a common epoch and not the observed values.

On the basis of purely kinematical considerations, an isochronous representation for angular diameters may be derived. If a signal leaves a receding galaxy at time t' when the galaxy is at distance Δ', then the distance Δ of the galaxy at time t when the signal is observed will be,

$$\Delta = \Delta' + cz(t-t') = \Delta'(1+z),$$

or in terms of the linear and angular diameters,

$$S/\theta = [S'/\theta'](1+z)$$

where S' and θ' are respectively the *observed* linear and angular diameters corresponding to epoch t' and S and θ are the linear diameters reduced to their values at epoch t.

If the linear diameter has remained unchanged during the light travel time $(t-t')$, then

$$\theta = \theta'(1+z)^{-1}.$$

This is the kinematic correction to angular diameters applicable to all cosmological models[8]. The above logarithmic relation may now be rewritten in terms of observed angular diameters θ'. Substituting for θ,

$$\log S = \log S' = \log\theta' + \log z - \log(1+z) + \log(c/H)$$

or

$$\log S' = \log\theta' - \log u + \log(c/H). \qquad (42.1)$$

The quantity $u = (1+z)/z$ will be designated the "synoptic redshift". Eq. (42.1) is derived from local kinematic considerations and does not include higher order redshift or evolutionary effects. It should, however,

[8] Hoyle, F.: Cosmological Tests of Gravitational Theories. In: *Proc. Int. School of Phys.*, vol. XX, pp. 141–173. New York: Academic Press 1962.

be applicable locally. For more distant galaxies, various cosmological models introduce modifications involving different powers of $(1 + z)$ into the diameter distance realtion. To be inclusive as possible, the tests for discretization among non-local samples should therefore be based on generalizations of the form,

$$\log S = \text{constant} + \log \theta + \log z + \sigma \log (1 + z) \tag{42.2}$$

where the exponent σ is to be empirically determined and compared with values predicted by various cosmological models.

The basic test equation of the discretization hypothesis may now be written by combining the logarithmic form of (37.1) with either (42.1), which is valid locally, or with the more general form (42.2), for which a cosmological parameter σ must be derived. The local test equation becomes

$$\log \theta' - \log u = \text{constant} + \tfrac{1}{2} \log n(n + 1) + \log l_i \tag{42.3}$$

where θ' is the observed angular diameter, $u = (1 + z)/z$, n is a positive integer and l_i is a physical parameter with dimension length. The general test equation becomes

$$\log \theta' + \log z + \sigma \log (1 + z) = \text{constant} + \tfrac{1}{2} \log n(n + 1) + \log l_i \tag{42.4}$$

where σ is to be determined.

In the special case of galaxies belonging to the same cluster, the test equation takes a particularly simple form. If the cluster membership of each member of the sample can be established by redshifts or some other suitable criterion, then the spread in distance among members of the sample should be small, and angular diameters become surrogate linear diameters. An estimate of the size of error introduced by this approximation may be made by assuming that the member galaxies of a cluster are distributed in depth along the line of sight by an amount equal to their linear distribution at right angles to the line of sight. If Δ is the distance to the cluster and Φ is its angular extension (measured in radians in the plane of the sky), then the relative distribution of depth is $\Phi \Delta / \Delta$. For example, if the Coma Cluster has a diameter of $2°$, its member galaxies will have a 3.5% fluctuation in relative distance. Whenever the relative depths are of the order of or less than the acceptable error in the diameter measurements, as will be true for the Coma Cluster and more distant clusters, then angular diameters uncorrected for distance may be tested for discretization effects directly. Thus for purposes of tests in clusters, the test equation may be written in the form

$$\log \theta' = \text{constant} + \tfrac{1}{2} \log n(n + 1) + \log l_i \tag{42.5}$$

where θ' is the observed angular diameter.

43. Discretization in Bright Galaxies

The difficulties in defining and measuring diameters have been such that it has not been possible to obtain uniform sets of measurements of sufficient accuracy and consistency to be useable either for cosmological investigations or discretization tests. For example, the Shapley Ames Catalogue[9] which lists a large sample of nearby low eccentricity ellipticals, turns out to be unsuitable for discretization tests because of the large percent of diameter measurements which suffer round-off effects. As discussed in the Prologue, several sets of similarly defined diameters (such as diameters defined by different isophotal intensity levels) are needed in order to establish definitively the existence of a proportionality range. It might seem that different sets of published diameters of the same galaxies might be substituted for this purpose. But unless the operational definitions used are similar this is not necessarily so. It will accordingly be necessary to use data that may be too limited to establish proportionality, but for which proportionality is inferred from the Hubble luminosity distribution law.

In this and the following section, the angular diameter measurements from two sets of data are compared with the predicted discretization sequences and a measure of the degree of fit between the observations and theory established for each sample. Statistical significance of the fits and some paradigmatic inferences are given to evaluate the results.

The first test sample consists of the bright EO galaxies from de Vaucouleurs' catalogue[10]. The data from the de Vaucouleurs catalogue consists of all galaxies classified by the authors as EO for which redshifts and diameters are available. In Table 6, a summary of the de Vaucouleurs' data and various calculated quantities which are required for the subsequent discussion are tabulated for 31 galaxies. The first column of Table 6 gives the NGC numbers of the galaxies with the parentheses indicating galaxies whose classification as EO's may be uncertain but whose major and minor axes are given as equal. The second column gives the common logarithm of the apparent major axis or diameter in tenths of a minute of arc. The third column gives the velocity corrected for galactic rotation, V_c. The fourth column is the common logarithm of the synoptic velocity or synoptic redshift. The seventh column gives the $B-V$ colors reduced to the diameter D and corrected for galactic absorption and redshift. The angular diameters are derived statistically from published values of micrometric diameters by several observers.

---

[9] Shapley, H., Ames, A.: *Ann. Harvard Obs. 88,* 2 (1932).

[10] Vaucouleurs, A. and G. de: *Reference Catalogue of Bright Galaxies.* Austin: University of Texas Press 1964.

The redshifts are either Humason-Mayall-Sandage values[11] or recent determinations by A. and G. de Vaucouleurs. Details of the diameter measures, the magnitude and color systems and the corrections are given in the de Vaucouleurs Catalogue. The tabular values in Table 6

Table 6. *Diameters and colors of bright galaxies*

NGC	$\log D$	V_c	$\log u$	P	δP	Color
83	0.94	6745	1.658	1.203	0.01	0.91
382	0.81	5354	1.756	1.030	0.01	–
596	1.28	2097	2.158	1.321	0.02	0.78
741	1.15	5636	1.734	1.379	0.01	0.90
(751)	0.88	5291	1.761	1.097	0.01	0.75
1374	1.01	1136	2.423	0.933	0.03	–
1379	1.05	1303	2.364	0.999	0.02	–
1399	1.26	1311	2.363	1.210	0.02	0.73
1407	1.16	1707	2.246	1.162	0.02	0.95
1889	0.65	2334	2.112	0.711	0.02	0.77
2673	0.89	3669	1.918	1.038	0.01	0.65
(2694)	0.90	5165	1.771	1.113	0.01	0.89
3348	1.02	3010	2.003	1.130	0.01	–
4283	0.94	1078	2.446	0.853	0.03	0.86
4339	1.15	1173	2.410	1.079	0.03	0.83
4458	1.05	311	2.985	0.723	0.08	0.80
4552	1.30	195	3.187	0.884	0.10	0.91
4636	1.48	629	2.679	1.330	0.04	0.85
4782	0.96	3858	1.896	1.117	0.01	–
4783	0.97	4527	1.828	1.158	0.01	–
(4827)	1.04	7657	1.604	1.327	0.01	0.81
4915	0.96	3034	1.999	1.072	0.01	0.79
(4926)	1.00	7673	1.603	1.288	0.01	0.96
5061	1.18	1888	2.204	1.200	0.02	–
5173	1.00	2508	2.081	1.075	0.02	–
5812	0.96	2040	2.170	0.996	0.02	0.88
5846	1.32	1784	2.228	1.330	0.02	0.89
5898	0.88	2231	2.132	0.932	0.02	0.74
5930	1.09	2868	2.024	1.190	0.01	–
(5953)	0.99	2188	2,140	1.039	0.02	–
7507	1.05	1684	2.253	1.049	0.02	0.81

NGC is catalogue number.
D is major axis in 0.1 min arc.
V_c is velocity corrected for galactic rotation in km/sec.
u is $(c + V_c)/V_c = (1 + z)/z$.
$P = 1 + \log D - \sigma \log u$ for $\sigma = 4/9$.
δP is measure of relative uncertainty.
Color is $B - V$ corrected for size, latitude, and redshift

[11] Humason, M. L., Mayall, U., Sandage, A. R.: *Astron. J. 61*, 97 (1956).

are carried to a larger number of places than the precision in the individual measurements warrants. This is done throughout in order to minimize round-off effects (discretization must not be injected into the data by rounding off).

The errors in $\log D$ are not known, but if they may be infereed from the internal errors of the various measurements used in arriving at $\log D$'s, a value $\leqq \pm 0.01$ (i.e., 3%) may be taken. This error will be assumed throughout the range of $\log D$ for the galaxies reported in Table 6. The error in the redshift is more or less independent of the size of the displacement resulting in a large relative error for nearby objects and a small relative error for distant objects. A constant error of ± 70 km/sec[12] is assumed for V_c. The errors in $\log u$ range from almost 0.2 for the nearest galaxies in Table 6 to about 0.004 for the most distant galaxy. The redshift and diameter errors are equal at velocities of about 3000 km/sec.

Provided that suitable corrections have been made to the D's, the linear diameters, S, given by Eqs. (42.3) or (42.4) should be metric diameters suitable for testing the discretization hypothesis. The first discretization tests on the data of Table 6 were run on a set of 31 linear diameters.

$$\{A\} = \{\log D + \log V_c\},$$

using a method to be described in detail later. No statistically significant discretization sequences were found for these A's. However, using a set of 31 linear diameters,

$$\{B\} = \{\log D - \log u\},$$

three discretization sequences, fitting 29 of the 31 members of the sample, were found. The sensitivity of the fits to the factor $(1 + z)$, a factor independent of the discretization parameters, increases the level of confidence in the reality of the sequences and in the uniformity of the data. It also permits refined estimates of the internal errors.

The values of $\log D$ "compacted" on the eigen sequence (i.e., $\log D - \frac{1}{2}\log[n(n+1)]$ using a suitable value of n from 1 to 8) are shown in Fig. 33. The significant sets corresponding to three values of the dimensional parameters l_i are shown as crosses, circles, or squares in Fig. 33. The slope of the compacted array leads to

$$\log D - \tfrac{4}{9}\log u$$

---

[12] The mean value for nearby galaxies reported in the Humason-Mayall-Sandage catalogue, *Astron. J. 61*, 97 (1956).

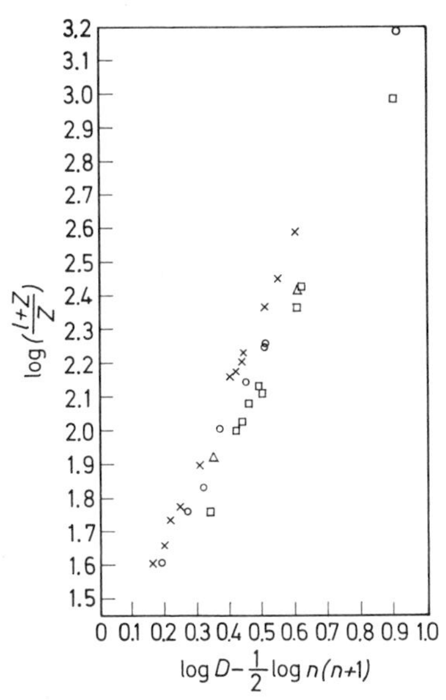

Fig. 33. Compacted bright galaxy discretization classes

as the function consistent with the theoretical discrete eigen functions. Accordingly, the quantities,

$$P = 1 + \log D - \tfrac{4}{9} \log u$$

are calculated and given in Table 6 along with δP's which are not internal errors but measures of relative uncertainties of the P's. The factor $\frac{4}{9}$ is unexpected, but P may still be taken as a "quasi diameter" representing the size of the galaxy since $\log D$ + constant and $\log u$ are dimensionless. Whether the "$\frac{4}{9}$" is peculiar to the data sample reflecting a relation between the operational and metric diameters, cosmological, or attributable to *bi-discretization*[13] is not at once clear. The question of how the quantity P is related to the quantities theoretically predicted to occur in sizes consistent with eigen sequences of the form $[n(n+1)]^{\frac{1}{2}}$ is unanswerable for reasons discussed in the Prologue. For testing of P's, the formulation of the discretization hypothesis given in Section 42 may be modified to state that there should exist a positive integer n such that for each P of the set,

[13] Whenever there is two parameter discretization, there may be multi-discretizations. This is obvious in the simple case of two equi-interval discretizations. If there exist equally spaced rows and columns, then there exist equally spaced diagonals with slopes of 1:1, 1:2, 1:3. etc. Multi-discretizations may also arise in certain cases of non-equi-interval bi-discretization.

$$P - \tfrac{1}{2}\log\left[n(n+1)\right] = \text{constant} + \log G_i \tag{43.1}$$

to within a specified deviation with the same proviso that, if the n's are to be discriminable, then G_i must assume only a limited number of values.

The analysis of the data sample for the presence of discretization sequences was made by comparing observed and theoretical *difference tables*. In order to isolate and identify any integers that may satisfy the discretization hypothesis, the unknown constants appearing in Eq. (43.1) may be eliminated by comparing the differences between P's with the set of differences,

$$Q_{ij} = \tfrac{1}{2}\log\left[i(i+1)\right] - \tfrac{1}{2}\log\left[j(j+1)\right]$$

for $i \geqq j$. First, the discrete values of P are ordered (values of P differing by less than the internal error may be combined and averaged), then a table of all possible differences between values of P constructed. This table is compared with a similarly ordered table of all possible Q_{ij}'s. The coincidence in value and relative position of patterns of three or more terms provides an identification of 3 or more branch numbers on which a discretization sequence may be based. Statistically significant sequences usually require at least four branches with the coincidence of six terms in the difference tables. The most important differences are those between the smallest values of P which presumably should correspond to the differences between the lowest values of n in $\frac{1}{2}\log\left[n(n+1)\right]$. Because of smaller separations, fits for larger values of n have less significance and should be given little weight unless they are part of a sequence which also fits the lower values of n.

Several of the P's in Table 6 are seen to cluster, differing by only two or three parts per thousand. When the differences are less than the estimated internal errors (about $0.1\,\delta P$), the mean P of a clustering is substituted for the separate values. The EO bright galaxy sample of 31 thus maps onto 21 distinct values. Comparison of a difference table of P's with a table of Q_{ij}'s reveals that 11 galaxies of the sample occupy six branches of a sequence possessing a value of

$$P - \tfrac{1}{2}\log\left[n(n+1)\right] = G$$

equal to 0.459 for $n = 2, 3, 4, 5, 7$, and 8. Table 7 shows the comparison of the differences of P's and Q_{ij}'s of this group of galaxies. The small size of the residuals and percent errors (given in Table 8) together with the large number of branches affords positive identification of the set.

The remaining galaxies of the sample may be identified in a similar way with sets of eigen sequences having other values of G, the G's being the analogues for P's of the log l_i's of the B's. A total of six sequences were found that together spanned 29 of the sample of 31 to within an

Table 7. *Difference table bright galaxies (Sequence 3)*[a]

P_i / P_j	0.853 ($i=2$)	0.998 ($i=3$)	1.115 ($i=4$)	1.202 ($i=5$)	1.329 ($i=7$)	1.379 ($i=8$)
0.853 ($j=2$)	0	0.145 [0.151]	0.262 [0.261]	0.349 [0.350]	0.476 [0.485]	0.526 [0.540]
0.998 ($j=3$)		0	0.117 [0.111]	0.204 [0.199]	0.331 [0.335]	0.381 [0.389]
1.115 ($j=4$)			0	0.087 [0.088]	0.214 [0.224]	0.264 [0.278]
1.202 ($j=5$)				0	0.127 [0.136]	0.177 [0.190]
1.329 ($j=7$)	$P = 1 + \log D - 4/9 \log u$				0	0.050 [0.055]
1.379 ($j=8$)	$Q_{ij} = 1/2 \log [i(i+1)] - 1/2 \log [j(j+1)]$					0

[a] Main entries are $P_i - P_j$. Bracketed entries are Q_{ij}.

Table 8. *Residuals (Sequence 3) difference table*[a]

P_i / P_j	0.853 ($i=2$)	0.998 ($i=3$)	1.115 ($i=4$)	1.202 ($i=5$)	1.329 ($i=7$)	1.379 ($i=8$)
0.853 ($j=2$)	0	0.006 [0.040]	−0.001 [0.004]	0.001 [0.002]	0.009 [0.019]	0.014 [0.026]
0.998 ($j=3$)		0	−0.006 [0.054]	−0.005 [0.025]	0.004 [0.012]	0.008 [0.021]
1.115 ($j=4$)			0	0.001 [0.011]	0.010 [0.045]	0.014 [0.050]
1.202 ($j=5$)				0	0.009 [0.066]	0.013 [0.068]
1.329 ($j=7$)	$P = 1 + \log D - 4/9 \log u$				0	0.005 [0.091]
1.379 ($j=8$)	$Q_{ij} = 1/2 \log [i(i+1)] - 1/2 \log [j(j+1)]$					0

[a] Main entries are $Q_{ij} - (P_i - P_j)$. Bracketed entries are $[Q_{ij} - (P_i - P_j)]/Q_{ij}$.

allowed error of 5%[14]. However, only three of these sequences (Nos. 3, 4 and 5) were sufficiently rich to be considered well determined.

The six sequences in order of descending value of G are listed in Table 9. In the left column are listed in order of increasing size, the 21 distinct values of P followed by a parenthetical number indicating cases

[14] One of the two non-member galaxies (NGC 4458) had an excessive $\delta P = 0.08$. The other galaxy (NGC 7507) may have been of higher eccentricity and a discernable member of the orientation contamination.

Table 9. *Bright galaxy discretization sequences*[a]

Sequence / P's	1	2	3	4	5	6
0.711	0.560 $[n = 1]$	–	–	–	–	–
(0.723)	–	–	–	–	–	–
(0.853)	–	–	0.464 $[n = 2]$	–	–	–
(0.884)	–	0.495 $[n = 2]$	–	–	–	–
0.933 (2)	–	–	–	–	0.393 $[n = 3]$	–
0.998 (2)	–	–	0.458 $[n = 3]$	–	–	–
1.030	–	–	–	–	–	0.379 $[n = 4]$
1.039 (2)	–	0.499 $[n = 3]$	–	–	0.388 $[n = 4]$	–
1.049	–	–	–	–	–	–
1.075 (3)	–	–	–	0.424 $[n = 4]$	–	–
1.097	0.557 $[n = 3]$	–	–	–	–	–
1.115 (2)	–	–	0.464 $[n = 4]$	–	–	0.376 $[n = 5]$
1.130	–	–	–	–	0.391 $[n = 5]$	–
1.160 (2)	–	–	–	0.421 $[n = 5]$	–	–
1.190	–	–	–	–	–	0.378 $[n = 6]$
1.202 (2)	–	–	0.463 $[n = 5]$	–	0.390 $[n = 6]$	–
1.210	0.559 $[n = 4]$	–	–	–	–	–
1.288	0.549 $[n = 5]$	–	–	0.414 $[n = 7]$	–	–
1.321	–	–	–	–	0.392 $[n = 8]$	–
1.329 (3)	–	–	0.455 $[n = 7]$	–	–	–
1.379	0.567 $[n = 6]$	0.505 $[n = 7]$	0.450 $[n = 8]$	–	–	–
Maximum sequence membership:	5	4	11	6	8	4
Mean $G = \bar{G}$:	0.558	0.500	0.459	0.421	0.391	0.377

[a] Entries are calculated G's. Bracketed entries are branch numbers. $P = 1 + \log D - 4/9 \log u$. $G = P - 1/2 \log [n(n + 1)]$.

of multiple occupancy of the value. For each sequence, the values of the branch numbers, n, and the derived values of G are given. At the bottom of the table are listed for each sequence, the mean G's and the number of galaxies fitted by the sequence. Monte Carlo attempts to generate equal fits to an eigen sequence based on uniform density random numbers of the same range as the P's failed in 100 tests on the RAND 7090 computer.

It is seen that the fits are not unique. Within the accuracies of the P's, seven of the galaxies may be assigned equally well to two sequences, and one large galaxy (NGC 741) to three sequences. (Interestingly, however, six of the eight galaxies of ambiguous sequence membership also are associated with multiple occupancy of a P value possibly indicating that density of occupancy of a P value may be attributable to overlap of sequence values.) Definitive assignments are not possible without higher precision data or some additional criterion of sequence discrimination.

A possible clue to discrimination in the cases of ambiguous sequence membership may lie in the observed *colors* of galaxies. In Table 10, all of the sequences are listed with the colors of the galaxies when available. The unambiguous members of each sequence serve to define a mean color class for the sequence. For example, in Sequence 1, a mean color of 0.75 is suggested by the unambiguous membership for those values of P corresponding to $n = 1$, $n = 3$, and $n = 4$. On the basis of comparing colors of ambiguous galaxies with mean sequence color, the following assignments of ambiguous galaxies seem appropriate:

NGC 4926 ($P = 1.288$) to Sequence 4
NGC 741 ($P = 1.379$) to Sequence 2[15]
NGC 2673 ($P = 1.038$) to Sequence 5
NGC 2694 ($P = 1.113$) to Sequence 3
NGC 83 ($P = 1.203$) to Sequence 3.

Lacking colors, the proper assignment of NGC 5953 ($P = 1.039$), NGC 4782 ($P = 1.117$), and NGC 5061 ($P = 1.200$) remains undecidable. With these assignments, the mean colors are those given below each sequence in Table 10.

The discretization sequences are therefore not only discriminated on the basis of the G's, but also in accordance with color, an independent physical observable. Furthermore, the set $\{\bar{G}\}$ itself seems to possess a

[15] The ambiquity in the assignment of NGC 741 ($P = 1.379$) can be reduced but not resolved. NGC 741 may be assigned to either Sequence 2 or 3, but not to Sequence 1. The choice is made for Sequence 2 on the basis of an apparent trend in color with branch number that NGC 741 appears to violate if placed in Sequence 3. However, the assignment to either Sequence 2 or 3 does not significantly change the mean color of either class.

Table 10. *Mean colors of bright galaxy discretization sequences*

Sequence 1

NGC	P	n	Color	Alternate[a]
1889	0.711	1	0.77	–
751	1.097	3	0.75	–
1399	1.210	4	0.73	–
4926	1.288	5	0.96	4
741	1.379	6	0.90	2, 3
		mean:	0.75	

Sequence 2

NGC	P	n	Color	Alternate[a]
4552	0.884	2	0.91	–
2673	1.038	3	0.65	5
5953	1.039	3	–	5
741	1.379	7	0.90	1, 3
		mean:	0.91	

Sequence 3

NGC	P	n	Color	Alternate[a]
4283	0.853	2	0.86	–
5812	0.996	3	0.88	–
1379	0.999	3	–	–
2694	1.113	4	0.89	6
4782	1.117	4	–	6
5061	1.200	5	–	5
83	1.203	5	0.91	5
4827	1.327	7	0.81	–
5846	1.330	7	0.89	–
4636	1.330	7	0.85	–
741	1.379	8	0.90	1, 2
		mean:	0.86	

Sequence 4

NGC	P	n	Color	Alternate[a]
4915	1.072	4	0.79	–
5173	1.075	4	–	–
4339	1.079	4	0.83	–
4783	1.158	5	–	–
1407	1.162	5	0.95	–
4926	1.288	7	0.96	1
		mean:	0.84	

Sequence 5

NGC	P	n	Color	Alternate[a]
5898	0.932	3	0.74	–
1374	0.933	3	–	–
2673	1.038	4	0.65	2
5953	1.039	4	–	2
3348	1.130	5	–	–
5061	1.200	6	–	3
83	1.203	6	0.91	3
596	1.321	8	0.78	–
		mean:	0.72	

Sequence 6

NGC	P	n	Color	Alternate[a]
382	1.030	4	–	–
2694	1.113	5	0.89	3
4782	1.117	5	–	3
5930	1.190	6	–	–

[a] Numbers in this column refer to alternate sequences to which the galaxy may be assigned on the basis of discretization fits.

regular discretized structure. In a manner similar to the analysis of the set $\{P\}$ to detect discretization sequences based on the eigen sequence $[n(n+1)]^{\frac{1}{2}}$, the set $\{\bar{G}\}$ may be shown to fit the sequence $[v(v+1)]^{\frac{1}{3}}$ where v is a positive integer. In fact,

$$G + 1/3 \log [v(v+1)] = E, \quad \text{a constant}. \tag{43.2}$$

The values of v, G, and E are given in Table 11. Omitting the low weight sequence 6, the mean value of E turns out to be 1.043.

The existence of a set of color classes to which $E0$ galaxies may be assigned on the basis of a discretization parameter v complicates the original discretization hypothesis, but provides the condition, namely a limited number of l_i classes, under which discretization may be detected. However, the correlation of the $\bar{G}_v$ discretization sequences with color, like the appearance of the factor 4/9 in the synoptic redshifts, raises many questions. One hypothesis is that the parameter v may be an age parameter[16]. In this event, we may have corroboration of the hypothesis that cosmic bodies possess discrete epochs of creation as proposed in "sequences of little bangs" cosmologies[17].

From the well established evolutionary tracks of stars[18], the predicted color evolution of a galaxy would consist of gradual reddening as the main sequence ages, followed by a more pronounced reddening as larger numbers of stars enter the giant stage. The final stages, consistent with evolutionary star tracks, would see the galaxy again returning to a mean color that is much bluer. If the v parameter is given the interpretation of epoch of creation, the colors given in Table 11 march with

Table 11. *Color class discretization*

Sequence	v	$\bar{G}$	E	Mean color
1	5	0.558	1.050	0.75
2	6	0.500	1.041	0.91
3	7	0.459	1.042	0.86
4	8	0.421	1.040	0.84
5	9	0.391	1.042	0.72
(6)	(10)	(0.377)	(1.058)	–

the value of v in a consistent manner. The largest values of v are to be associated with the most recent epochs of creation and the youngest classes of galaxies. Proceeding to successively older classes, from $v = 9$, to 8, to ..., the predicted color evolution is seen in the change toward red (larger numbers) of the mean color classes in Table 11. The oldest class $v = 5$ has reached the age when evolution toward the blue again occurs. Still older unobserved classes $v = 4, 3, \ldots$ may be bluer still or even consist of dark galaxies. Quasars may fit on either end of this

[16] Wilson, A. G.: *Astron. J.* *69,* 153 (1964).

[17] Fowler, W. A.: In: *High Energy Physics and Nuclear Structure,* pp. 203 to 225. Amsterdam: North-Holland 1967.

[18] See for example, Burbidge, G. R., Burbidge, E. M.: In: *Handbuch der Physik,* Vol. 51, p. 134. Berlin-Göttingen-Heidelberg: Springer 1958.

evolutionary sequence, being new galaxies coming into existence from a "new bang" in a new epoch, or may be very old galaxies undergoing gravitational collapse, passing through a stage of evolution analogous to the passage to the white dwarf stage in stellar evolution.

In summary, it is not justified to conclude that the discretization hypothesis has been verified by the sample of diameters of bright $E0$ galaxies from the de Vaucouleurs catalogue. This is principally because of the very limited size of the sample and the interpretation difficulties introduced by the 4/9 power of the synoptic redshifts. However, several important results affirming the *consistency* of observation with the discretization hypothesis have emerged. In addition to Monte Carlo tests discounting the fits to chance, several paradigmatic points reinforce the discretization model. These include inferences from what did *not* fit as well as from what did fit. The set of diameters based on redshifts, z, showed no convincing sequences; while diameters based on the synoptic redshift $(1 + z)/z$ fell readily into three discretization sequences. Since the synoptic redshifts supposedly provide an isochronous uniform sample, this pair of results, one negative and one positive is supportive. The existence of a simple recipe for mapping the entire discretization structure on a single constant, first through compaction on the n or diameter sequence and then on the v or color sequence provides a satisfying consistency and generality to the discretization concept, even though the v classes were not predicted by theory. In the spirit of the epistemological viewpoint of the Prologue, we may conclude that, whereas no satisfactory answer to the question "are the diameters of bright $E0$ galaxies discretized" can be given, the *organization* of the diameter data using the "discretization operation" leads to an economy of representation of the data and when unique to a useful classification schema.

44. Discretization in Cluster Galaxies

The original observational evidence for discretization was found in 1950 in the comparison of microdensitometer tracings of galaxies in the Coma Cluster, the Ursa Major I Cluster, the Corona Borealis Cluster and the Boötes Cluster. Several tracings of $E0$ galaxies in each cluster were nearly identical when superimposed. Intercluster comparisons of the angular diameters of these congruent galaxies showed the existence of at least three distinct families of sizes having members in each cluster.

This study was repeated in 1962 on a set of homogeneous plates made by Humason with the 200 inch telescope of the Ursa Major II and Coma B Clusters and the above four clusters. The second set of measurements of apparent $E0$ galaxies, both micrometric and from microdensitometer tracings, showed the same three major classes of galaxy size marked s, t, and u in Fig. 34.

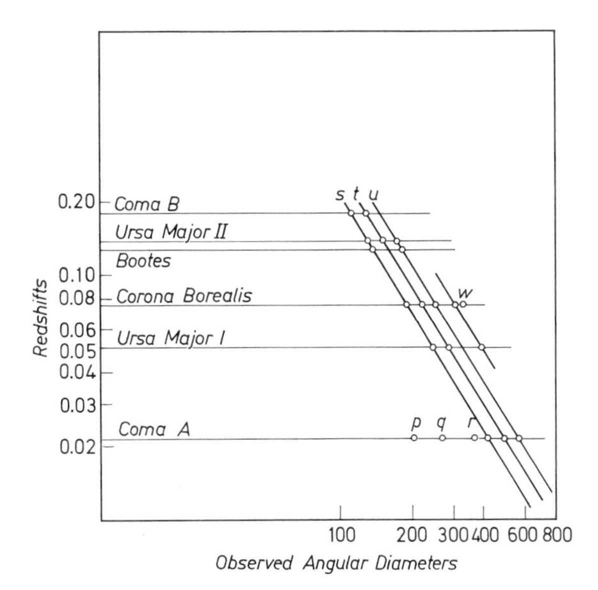

Fig. 34. Size classes of *EO* galaxies

While evidence for discrete size classes of galaxies had been uncovered, no systematic relation between the diameters in the various size classes had been found. The theoretical discretization hypothesis predicting that diameters belonging to different size classes should be proportional to $\sqrt{n(n+1)}$ made available, in addition to inter-cluster comparisons, a second method of grouping the apparent size coincidences and enlarging the discretized structure.

Using the difference table methods described in Section 43, a sample of 30 *EO* galaxies distributed among the six clusters were compacted according to the sequence $\sqrt{n(n+1)}$. Within a cluster, the angular diameters, D, were assumed proportional to the linear diameters in accordance with Eq. (42.5). Quantities,

$$G_0 = \log D - 1/2 \log [n(n+1)]$$

were evaluated and the set of mean $\bar{G}_0$'s compared from cluster to cluster making corrections for redshifts. No congruences between $\bar{G}_0 + \log \bar{z}$ or $\bar{G}_0 - \log \bar{u}$ were found (where $\bar{z}$ is the mean redshift for the cluster and $\bar{u} = (1+\bar{z})/\bar{z}$) as would be expected from equivalent diameter classes. However, the quantities $\bar{G}_0 - \frac{2}{3} \log \bar{u}$ took on the same set of values from cluster to cluster as may be seen in column five of Table 12. Hence the quantities

$$\log D - \tfrac{1}{2} \log [n(n+1)] - \tfrac{2}{3} \log \bar{u} = F \qquad (44.1)$$

assume only a limited number of values, the necessary condition for discretization detection. (This is similar to the occurence of a finite number

Table 12. *Discretization classes in clusters*

Cluster	$\log \bar{u}$	Sequences[a]	G_0	$G_0 - \frac{2}{3}\log \bar{u}$
Coma	1.678	7	1.768	0.649
		3	1.694	0.575
		4	1.640	0.521
Ursa Major I	1.315	4	1.454	0.577
Corona Borealis	1.172	5	1.416	0.635
		6	1.358	0.577
		4	1.305	0.524
		6	1.267	0.486
Boötes	0.936	(3)	1.262	0.638
		(3)	1.040	0.416
Ursa Major II	0.921	(3)	1.143	0.529
		5	1.026	0.412

[a] The number of galaxies in the sequence.
$\bar{u} = (1 + \bar{z})/\bar{z}$ $G_0 = \log D - \frac{1}{2}\log[n(n+1)]$

of values of the quantity, $\log D - \frac{1}{2}\log[n(n+1)] - \frac{4}{9}\log u$, found in Section 43.) The set $\{F\}$ may in turn be compacted on the sequence $\frac{1}{3}\log[v(v+1)]$. In other words, there exist values of v, such that the quantities, $F + \frac{1}{3}\log[v(v+1)]$, take on only a single value. In Table 13 are listed all of the values of

Table 13. *Compacted cluster discretization classes*

Cluster	Values of F[a]	$\bar{F}$	v	$F + \frac{1}{3}\log[v(v+1)]$
Coma	0.649	(0.649)	(4)	(1.083)
Boötes	0.638	0.637	4	1.072
Cor Bor	0.635			1.069
U.M.I.	0.577			1.069
Cor Bor	0.577	0.577	5	1.069
Coma	0.575			1.067
U.M.II	0.529			1.070
Cor Bor	0.524	0.525	6	1.065
Coma	0.521			1.067
Cor Bor	0.486	0.486	7	1.069
Boötes	0.416	0.414	9	1.067
U.M.II	0.412			1.063
			Mean	1.0688

[a] $F = G_0 - \frac{2}{3}\log \bar{u} = \log D - \frac{1}{2}\log[n(n+1)] - \frac{2}{3}\log \bar{u}$.

$$F = \bar{G}_0 - \tfrac{2}{3}\log \bar{u},$$

together with the v's and the derived values of the single constant.

Except for the largest F sequence of the Coma Cluster, the remaining eleven sequences closely fit five values that readily compact with values of v of 4, 5, 6, 7, and 9 to the constant $C_0 = 1.069$. It follows that for the sample of EO and near EO galaxies in the five clusters (the statistical weight of the Coma B data being too low to include), the angular diameters D in each cluster have values given by,

$$\log D - \tfrac{2}{3}\log \bar{u} = C_0 + \tfrac{1}{2}\log[n(n+1)] - \tfrac{1}{3}\log[v(v+1)] \quad (44.2)$$

where $\bar{u}$ is the mean cluster redshift and $C_0 = 1.069$. From Eqs. (43.1) and (43.2), the equation corresponding to Eq. (44.2) for bright galaxies is

$$\log D - \tfrac{4}{9}\log u = (E-1) + \tfrac{1}{2}\log[n(n+1)] - \tfrac{1}{3}\log[v(v+1)]$$

where the value of E from Table 11 is 1.043. The cluster galaxies differ from the bright galaxies in the value of the constant and in the coefficient of the logarithm of the synoptic redshifts. The values of v found for the bright galaxies were 5, 6, 7, 8, 9 (and possibly 10); for the cluster galaxies $v = 4, 5, 6, 7$, and 9. There is insufficient color data for cluster galaxies to derive mean colors for the cluster v-classes. The effects of redshifts on color would have to be taken into account for intercluster comparisons.

The third investigation of discretization effects in cluster galaxies employed a new technique. A method of "stellarizing" galactic images was developed by W. C. Miller, George Kocher and A. Wilson. Plates were successively copied on high contrast emulsions under carefully controlled conditions resulting in the steepening of the galactic profiles until they presented hard stellar-like images. Comparison of microphotometric tracings of original and high contrast images showed that an increase in density gradient by a factor of four or better was effected in a single copy operation. The plate material for the third cluster discretization test was a set of high contrast copies of homogeneous plates (103*a*0, 30 min exposure) of eight clusters made at the prime focus of the 200 in. telescope by Sandage under uniform conditions of sky and seeing. The major axes of 130 low eccentricity stellarized images in the eight clusters were each measured five times with a micrometer. The values of d, the image diameter, ranged from 3.125 to 0.130 mm, which is equivalent to 34.4 to 1.4 sec of arc. Deviations from $\bar{d}$, the mean values of d, showed that the internal error of measurement was proportional to $\bar{d}$ with the r.m.s. error being about $0.007\,\bar{d}$. When small star images on the same high contrast copy plates were measured, values of 0.020 sec of arc were found. This value implied an increment in d due to optical imperfections and photographic irradiation as yet uncorrected. The possibility exists that the diameters measured in different clusters on different plates may be based on slightly different isophotes. Transfer plates photographing four different clusters in different quadrants of one plate were taken to calibrate such errors. The steep density gradient on the high contrast copies tends to reduce errors in d from this source. An estimate of a bound on this type of error can be made for "unstellarized" images from the Vaucouleurs formula[19] for the surface brightness in elliptical galaxies,

$$\log B(r) = \text{const} - 3.33 \log (r/r_e)^{\frac{1}{4}} .$$

It can be shown that an error of 0.2 or 0.08 in $\log B(r)$ corresponds to an error in the diameter d, of less than $0.1\, d$ (or 0.03 in $\log d$).

As in the earlier studies of cluster galaxies, the *linear* diameters were derived from the measured angular diameters, d_i, and the mean cluster redshifts $\bar{z}_j$, where $\bar{z}_j$ is the mean of *all* published redshifts for individual galaxies belonging to the jth cluster. In terms of the velocities, V,

$$\log S_{ij} = \log d_i + \log \bar{V}_j + \text{constant} \tag{44.3}$$

where S_{ij} is the linear diameter of the ith galaxy in the jth cluster and $\bar{z}_j = \bar{V}_j/c$. Estimated from the redshift dispersions in each cluster, the error in $\bar{z}_j$ is less than $0.02\, \bar{z}$. It follows that the relative errors *within* one cluster are the measurement errors $\Delta \log S_i$ which is less than ± 0.0085, while the maximum error between different clusters is given by

$$\Delta \log S_{ij} = \Delta \log d_i + 0.5\, \Delta \log B_j + \Delta \log \bar{z}_j \sim \pm 0.025.$$

Eliminating the photometric error in $\log B$ would reduce the error $\Delta \log S_{ij}$ to less than ± 0.01.

In the sample of 22 galaxies of apparent low eccentricity in the Coma Cluster, a sequence containing six galaxies compacts on the $\sqrt{n(n+1)}$ discretization sequence to a value of $K_1 = 3.5121$, fits to each galaxy being to within ± 0.005. Thus for six galaxies,

$$\log d_i + \log \bar{V}_j = K_1 + \tfrac{1}{2} \log [n(n+1)] \pm 0.005 \tag{44.4}$$

with $K_1 = 3.5121$. Compaction in the other seven clusters resulted in a total of 35 galaxies out of 108 forming sequences based on the *same* value of K_1, giving a total of 41 fits to the K_1 sequence out of the sample of 130. A second sequence was found in all of the clusters containing 34 galaxies based on the compaction constant, $K_2 = 3.6131$, and a third sequence was found containing 41 galaxies based on $K_3 = 3.6869$. Out of the 130 galaxies, 95 or about 70% belong to these three well defined sequences. Table 14 summarizes all of the fits to within ± 0.005.

---

[19] Vaucouleurs, G. de: In: *Handbuch der Physik,* Vol. LIII. Berlin-Göttingen-Heidelberg: Springer 1959.

(The total number of fits for the three values of K does not equal the total sample less the 35 that were not fitted because several galaxies fitted more than one sequence.)

Table 14. *Cluster galaxies discretization sequences of "stellarized" images*

Cluster	Values of n used with			Number of S_{ij}	
	$K_1 = 3.5121$	$K_2 = 3.6131$	$K_3 = 3.6869$	Total	Not fitted
Coma	5, 5, 8, 8, 8, 8	3, 7, 7, 12	1, 4, 6, 7, 8, 10, 16	22	6
2322 + 1425	4, 6, 8, 8, 12	10, 10	7, 9, 13	20	10
U.M.I	6, 7, 11, 15	13, 16	6, 6, 7, 7, 10, 11, 13	13	2
Cor Bor	8, 8, 8, 8, 10, 10, 10, 12, 14, 14, 16	6, 7, 8, 8, 8, 9, 10, 11, 11, 14, 15	3, 5, 6, 8, 10, 13, 13	29	8
Shane Cloud	9, 11, 12, 13, 15	7, 8, 8, 8, 9, 9, 12, 15	8, 9, 9, 10, 11, 14	19	5
Boötes	9, 10, 14, 17	8, 10	6, 7, 9, 15, 17	11	2
U.M.II	9, 13, 16	9, 9, 12, 12	6, 9, 14	11	2
Hydra	10, 12, 15	12	7, 9, 10	5	0

As in previous samples, the question of whether the K-class is correlated with any other physical observable arises. Unfortunately there is inadequate color data available to check the possibility of K-class being related to color. However, in the Coma cluster, it was found that the galaxies belonging to the different K-classes had different mean ellipticities. For class K_1, $\bar{E} = 1.16 \pm 1$; for K_2, $\bar{E} = 2.21 \pm 1$; and for K_3, $\bar{E} = 3.12 \pm 1$.

The statistical significance of the number and degree of fits of the data to the theoretical sequences is summarized in Tables 15 and 16. The range of observed $\log S_{ij}$ is from 3.80 to 4.95 or 1.15 and the largest value

Table 15. *Ratios of observed to random coincidences*

Tolerance	$N(\text{obs})/N(\text{prob})$	
δ	K_1	K_3
± 0.0050	2.10	1.60
± 0.0075	1.98	1.64
± 0.010	1.76	1.60
± 0.015	1.54	1.54
± 0.020	1.41	1.46

Table 16. *Probabilities of observed coincidences*

Tolerance δ	$N(\text{obs})/N_0$ for		$\hat{N}/N_0$ for		$\bar{N}(\text{prob})/N_0$	Standard score	Probability
	K_1	K_3	K_1	K_3		K_1	K_1
0.0050	0.3145	0.2405	0.21	0.24	0.15	5.25	All
0.0075	0.4348	0.3608	0.28	0.30	0.22	6.68	probabilities
0.010	0.5273	0.4810	0.36	0.37	0.30	5.43	are less
0.015	0.6475	0.6475	0.50	0.55	0.42	4.42	than
0.020	0.7308	0.7585	0.59	0.67	0.52	4.85	0.00001

The fit for K_1 is best, for K_3 is poorest.
$\bar{N}(\text{prob})$ is the mean number of fits in 25 random samples of size N_0.
$\hat{N}(\text{prob})$ is the maximum number of fits.
The standard score is $[N(\text{obs}) - \bar{N}(\text{prob})]/\sigma_n$ where σ_n is the dispersion.

of n is 17. Therefore, the expected probable number of coincidences, $N(\text{prob})$, between a set of N_0 random numbers having a uniform density distribution and the set of 17 prescribed values is

$$N(\text{prob}) = (2\delta)(17\,N_0)/1.15$$

where δ is the assumed error. (Actually the log S_{ij}'s are not uniformly distributed and for large δ and n, the intervals overlap, factors that must be taken into account.) The ratios of the observed number of "successes" (coincidences between log S_{ij} and the theoretical values to within $\pm\delta$) to the probable number of fits, $N(\text{obs})/N(\text{prob})$ is given for the best sequence K_1 and the worst sequence K_3 in Table 15. The values of observed successes greatly exceed probable successes, but fluctuations about the expected value of $N(\text{prob})$ could account for this. Accordingly, a Monte Carlo calculation was run 25 times to get the mean value of $N(\text{prob}) = \bar{N}(\text{prob})$, the maximum value $= \hat{N}(\text{prob})$ and the dispersion σ_n. In Table 16, the probabilities that the degree of fit achieved by the set $\{\log S_{ij}\}$ could be chance are given as a function of the tolerance, δ. It is seen that the highest number of σ_n's occurs near $\delta = \pm 0.008$, which tends to confirm both the error estimates and the validity of Eq (44.4).

Whether different values of K might provide a better fit or whether other discretization functions might provide better fits is an important question. Trials have shown that $[n(n+1)]^b$ does not fit nearly so well with $b = \pm 1$ or $-\frac{1}{2}$ as with $+\frac{1}{2}$. It is important, however, that the values of K were derived in the Coma Cluster and were successful in *predicting* the S_{ij}'s in the other seven clusters. Whether or not the particular discretization constants and functions employed provide the best possible fits, they are very successful in describing a large percentage of the observational data.

B. Redshift Discretizations

The empirical results of Part A are not only of interest because of their consistency with the theoretical discretization sequence and the support they lend to the usefulness of discrete formulations of various macro distributions, but also because of their implications and prediction of other discrete relations that may be checked observationally. It is paradigmatic inferences, more than the statistical results that support the importance of discrete formulations of macro phenomena.

In Part B, Section 45 will lead from the results of Part A into a new type of discretization which was not predicted by the basic theory. In Section 46, additional empirical results supporting the inferences of Section 45 are introduced. In Section 47, recently published discretization results based on new data will be summarized. This new type of discretization is, however, consistent with the discrete modes of inhomogeneity predicted by the prestructure approximation of conformally homogeneous world models (see Section 9).

45. Prediction of Discretized Redshifts

The extent of the empirical results given in Sections 43 and 44 have brought macro discretizations to the point where a more formal approach can be introduced. Accordingly, this section will be given to formalizing and structuring these results. We begin by defining more precisely an operation that has been repeatedly used on the bright galaxy and cluster galaxy data. This is the operation of *compaction.*

Compaction is a many to one mapping operation. It permits a set of measurements to be placed in correspondence with the integers and identifies the existence of subsets. Compaction is performed *with respect to* a specified arithmetic operation (e. g., division, exponentiation, etc.) and "through" a specified compaction sequence, $E(i)$, such as $[i(i+1)]^{\frac{1}{2}}$ with $i = 1, 2, \ldots$ The compaction algorithm consists of replacing every member P_i of a set of quantities $\{P\}$ by a single quantity P_0 whenever there exists an integer i such that $P_i * E(i) = P_0$, where $*$ is the specified arithmetic operation. The unmapped members of the set $\{P\}$ are called the *residual*; the quantity P_0 is called the *base.* A compactable set is one for which the residual is a null set. Formally, *A set $\{P\}$ is said to be compactable to the base P_0 through the compaction sequence $E(i)$ with respect to the operation $*$, if there exists an integer, $i = 1, 2, \ldots$ such that for each member P_i of the set,*

$$P_i * E(i) = P_0 .$$

In practice a *tolerance*, δ, is usually introduced so that the set is compactable if for each P_i,

$$P_i \dot{*} E(i) = P_0 \pm \delta .$$

As an example, the results of Section 43 may be summarized as follows: For the bright galaxies, a set of observed diameters, $D_{n,v}$ and their synoptic redshifts $u_{n,v}$ were used to form the set

$$\{P_{n,v}\} = \{1 + \log D_{n,v} - \tfrac{4}{9} \log u_{n,v}\} .$$

The subscripts indicating that the set is a two parameter set. This set was found to be compactable through the sequence

$$E_2(n) = \tfrac{1}{2} \log [n(n+1)]$$

with respect to subtraction and then compactable through the sequence

$$E_3(v) = \tfrac{1}{3} \log [v(v+1)]$$

with respect to addition, i.e.,

$$\{\{P_{n,v}\} \dot{-} E_2(n)\} \dot{+} E_3(v) = 1.043 . \tag{45.1}$$

The compaction notation, $\{A_i\} \dot{-} E(i)$, means that the appropriate member of the sequence $E(i)$ is to be subtracted from each member of the set $\{A_i\}$.

In the case of the cluster galaxies, we may write

$$\{\log D_{n,v,j}\} \dot{-} E_2(n) = \{\log D_{0,v,j}\} \tag{45.2}$$

i.e., the logarithms of the set of measured angular diameters $\{D_{n,v,j}\}$ in cluster j was compacted with respect to subtraction through the sequence $E_2(n)$ generating a set of bases $\{\log D_{0,v,j}\}$ parameterized in each cluster, j, by v. (The set of bases was designated G_0 in Table 12.) The quantity $\tfrac{2}{3} \log \bar{u}_j$ for each cluster j was then subtracted from each of the bases, but it was seen that the resulting set was independent of the cluster j, i.e.,

$$\{\log D_{0,v,j} - \tfrac{2}{3} \log \bar{u}_j\} = \{\log F_{0,v}\} . \tag{45.3}$$

Finally the set $\{F_{0,v}\}$ was compacted with respect to addition through the sequence $E_3(v)$,

$$\{\log F_{0,v}\} \dot{+} E_3(v) = C_0 = 1.0688 . \tag{45.4}$$

These statements formally summarize the procedures whose results are given in Tables 12 and 13.

It is possible to commute the order of performing the above operations, although from the statistical point of view, establishing v-bases within each cluster is much weaker than establishing them over the entire sample. Following (45.2), we may first compact within each cluster through $E_3(v)$,

$$\{\log D_{0,v,j}\} \dot{+} E(v) = \{\log D_{0,j}\} \tag{45.5}$$

then substract the appropriate quantity $\frac{2}{3}\log\bar{u}_j$,

$$\{\log D_{0,j} - \tfrac{2}{3}\log\bar{u}_j\} = C_0 = 1.0688\,. \tag{45.6}$$

Hence from Eqs. (45.3), (45.4), (45.5), and (45.6), we have:

$$C_0 = \{\log D_{0,v,j} - \tfrac{2}{3}\log\bar{u}_j\} \dot{+} E_3(v) = \{\{\log D_{0,v,j}\} \dot{+} E_3(v)\} - \{\tfrac{2}{3}\log\bar{u}_j\} \tag{45.7}$$

where the notation in the right member means subtracting the corresponding member of the set $\{\frac{2}{3}\log\bar{u}_j\}$ from the compacted set $\{\{\log D_{0,v,j}\} \dot{+} E_3(v)\}$. From Eqs. (45.3) and (45.6), it is seen that the set of synoptic redshifts behaves like a non-integral discretization sequence in the sense that subtraction of the quantity $\frac{2}{3}\log\bar{u}_j$ from either the set of bases $\{\log D_{0,v,j}\}$ or $\{\log D_{0,j}\}$ creates a set independent of j. From Eq. (45.7) it is seen that the substraction operation with $\frac{2}{3}\log\,\bar{u}_j$ is commutative with the compaction operation $\dot{+}\,E_3(v)$. The quantity $\frac{2}{3}\log\bar{u}_j$ could not have these properties unless the mean synoptic redshifts of the clusters possessed a regular structure and were themselves discretized according to some eigen sequence. We therefore infer that the set $\{\log\bar{u}_j\}$ for clusters of galaxies forms a compactable set. However, the algebra of compaction sets is not well enough understood to predict the form of the compaction sequence.

46. Redshift Discretization of Clusters of Galaxies

The inference of Section 45 that mean synoptic redshifts, $\bar{u} = (1+\bar{z})/\bar{z}$ of clusters of galaxies (where $\bar{z} = (\sum z_i)/n$ is the mean redshift for a cluster) are discretized was investigated for the sample of five clusters listed in Table 12. The data suggest a compaction sequence of the form,

$$E_1(N) = \log\left[N(N+1)\right],$$

giving

$$\{\log\bar{u}_N\} \dot{+} E_1(N) = \log\bar{u}_0\,. \tag{46.1}$$

Eq. (46.1) was tested against 28 clusters for which published redshifts are available. It was found that there exist integers, N, such that the mean clusters redshifts of 20 of the 28 clusters fit Eq. (46.1) to within the estimated observational error (these are those clusters listed in Table 17 for which the parameter $M = 2$). Of the remaining clusters, integers N were found such that 7 fit Eq. (46.1) with a different base (those clusters for which $M = 4$ in Table 17). Hence the mean synoptic redshifts formed a two parameter set $\{\bar{u}_{N,M}\}$. It was found that all of the clusters formed a compactable set with respect to addition through the two compaction sequences $E_1(N)$ and $E_1(M)$, i.e.,

Table 17. *Mean cluster redshifts observed and calculated*

Cluster	Name	No. of red shifts	$\bar{V}$	$\log \bar{u}$	N	M	P	$P - \log \bar{u}$
....	Virgo	73	1136	2.4234	3	2	2.4191	−0.0043
0316 + 4121	Perseus	7	5435	1.7494	7	2	1.7500	0.0006
0123 − 0137	NGC 541	43	5439	1.7494	7	2	1.7500	0.0006
1257 + 2812	Coma	50	6432	1.6780	4	4	1.6745	−0.0035
1627 + 3937	Abell 2199	19	9028	1.5344	9	2	1.5440	0.0096
1603 + 1755	Hercules	15	10775	1.4597	10	2	1.4569	−0.0028
2308 + 0720	Pegasus II	3	12821	1.3874	11	2	1.3777	−0.0097
2322 + 1425		2	13187	1.3757	11	2	1.3777	0.0020
1145 + 5559	Ursa Major I	4	15269	1.3149	12	2	1.3052	−0.0097
0106 − 1536	Haufen A	2	15872	1.3012	12	2	1.3052	0.0040
1024 + 1039	Leo	1	19489	1.2147	7	4	1.2272	0.0125
1239 + 1853		2	21533	1.1741	14	2	1.1761	0.0020
1520 + 2754	Corona Borealis	8	21651	1.1719	14	2	1.1761	0.0042
0705 + 3506	Gemini	2	23366	1.1360	15	2	1.1181	−0.0179
0348 + 0613		1	25644	1.1038	15	2	1.1181	0.0143
1513 + 0433	Shane Cloud	1	28333	1.0640	16	2	1.0637	−0.0003
1431 + 3146	Boötes	2	39367	0.9356	10	4	0.9341	−0.0015
1055 + 5702	Ursa Major II	2	40860	0.9213	19	2	0.9185	−0.0028
1153 + 2341	Abell 1413	2	42784	0.9037	7	6	0.9049	0.0012
1534 + 3749		3	45951	0.8767	20	2	0.8751	−0.0016
0025 + 2223		2	47836	0.8616	11	4	0.8549	−0.0067
1228 + 1050		2	49514	0.8487	11	4	0.8549	0.0062
0138 + 1840		1	51908	0.8312	21	2	0.8337	0.0025
1309 − 0105		1	52362	0.8280	21	2	0.8337	0.0057
1304 + 3110	Coma B	1	54917	0.8104	21	2	0.8337	−0.0163
0925 + 2044		1	57498	0.7937	22	2	0.7941	0.0004
1253 + 4422		1	59382	0.7819	12	4	0.7824	0.0005
0855 + 0321	Hydra	3	60860	0.7730	12	4	0.7824	0.0094

$P = 4.2765 - \log M(M+1) - \log N(N+1)$; $\log \bar{u} = \log[(c + \bar{V})/\bar{V}]$.

$$\{\{\log \bar{u}_{,M}\} \dotplus E_1(N)\} \dotplus E_1(M) = ш \tag{46.2}$$

where ш is a constant that has a least square value of 4.2765 for the 28 clusters[20].

In Table 17, the first five columns are: (1) the cluster designation in terms of its 1950.0 position; (2) the common name for the cluster; (3) the number of individual redshifts available for the calculation of the mean; (4) the mean velocity for the cluster,

$$\bar{V} = \frac{c}{n} \sum_{i=1}^{n} z_i ,$$

[20] Wilson, A. G.: *Proc. Nat. Acad. Sci. 52,* 847 (1964).

where $z_i = (\delta\lambda/\lambda)_i$ is the measured spectral displacement; and (5) the logarithm of the synoptic redshift $\bar{u} = (c + \bar{V})/\bar{V}$. The source of all redshifts in Table 17 is the Humason-Mayall-Sandage Catalogue[21], except for the clusters; NGC 541, Coma Abell 2199 and Hercules which are taken from other references[22] Columns (6) and (7) are the values of M and N. The values of the function,

$$P(M,N) = ш - \log M(M+1) - \log N(N+1), \qquad (46.3)$$

called the *structural mean redshift*, are given in column (8) while the comparisons between the *observed mean redshift* $\bar{u}$ and the structural mean redshift are given in column (9). The values of $\log \bar{u}$ and of the residuals, $P - \log \bar{u}$, are given to more significant places than are consistent with the precision of the observations. This is done to avoid contamination of discretization by possible round-off effects. The residuals range from less than 0.003 (lowest meaningful value) to 0.018, corresponding to a discrepancy of 4%.

It is apparent that if ш were very large, we could find large integer values of M and N that would give values of $P(M,N)$ fitting the set of observations to within any prescribed residual. It is the low values of N that are the most stringent statistically. Accordingly, the integers, N, in Table 17 are selected as the smallest integers which provide as many distinct values over the range of $\log \bar{u}$ (i.e., 2.4234 to 0.7730) as there are distinct cluster redshifts. Note that $P(M,N)$ may assume the same value for different combinations of M and N, with the result that the identification of M and N is not always unique, as was also the case for diameters.

While the sequences for $M = 2$ and $M = 4$ each contain enough members to establish the statistical significance of Eq. (46.1) for two different constants, there is no justification in the present sample of redshifts, other than analogy, for writing of Eq. (46.2) with two integral parameters, M and N, and for assigning the unusually rich cluster Abell 1413 to a tentative $M = 6$ sequence.

Before investigating the statistical significance of the agreement between the observed and structural mean cluster redshifts, it is necessary to evaluate the estimated errors in the observed values, $\bar{u}$. The uncertainty in $\bar{V}$ or $\bar{u}$ depends on: (1) the errors in the individual redshifts, (2) the velocity dispersion in the cluster, and (3) the size of the sample

[21] *Astron. J. 61*, 97 (1956).

[22] NGC 541 from Zwicky, F., Humason, M. L.: *Astrophys. J. 139*, 269 (1964); Coma from Lovasich, J. L., Mayall, W., Neyman, J., Scott, E. L.: *Proc. 4th Berkeley Symp. Mathematical Statistics and Probability*, p. 187. Berkeley: University of California Press 1960; Abell 2199 from Minkowski, R.: *Astron. J. 66*, 558 (1961); and Hercules from Burbidge, G. R., Burbidge, E. M.: *Astrophys. J. 130*, 629 (1959).

of measured redshifts in the cluster, together with the extent to which the sample is contaminated by including redshifts of nonmember galaxies.

Compared with other cosmological observables, the errors in redshifts are small. The total errors in redshifts are more or less independent of their sizes, and their relative errors are consequently small except for nearby objects. If we may take the redshifts published in the Humason-Mayall-Sandage Catalogue as typical, then we may assume in accordance with their estimates that each measured velocity is within 175 km/sec of the true value, and over half are within 35 km/sec of the true value. Assuming the extreme bound of 175 km/sec for each redshift, the *relative* error in the individual redshifts is less than 3% (except for the nearby Virgo cluster). The relative error in the more distant cluster becomes very small, and the values are correct to the third decimal place in z. The dispersion σ_i, for individual redshifts, will be taken as 35 km/sec.

A much larger source of uncertainty in the *mean* redshifts of clusters is the velocity dispersion. The values of the means, $\bar{V}$, and the dispersions, σ_v, for all clusters for which three or more redshift measurements have been obtained are given in Table 18. The contribution of σ_i to the error in $\bar{V}$ is negligible compared with σ_v for all clusters except the Hydra cluster (where the effect of σ_i has been included). The remaining columns in Table 18 give the relative error,

$$\mathrm{d}u/u = \sigma_v/\bar{V}(1+\bar{z})\sqrt{n}\,;$$

the logarithmic errors,

$$\delta_u = \log(u+\mathrm{d}u) - \log(u)\,;$$

and the observed residuals, ρ, from Table 17.

For the clusters in Table 18, the magnitudes of the estimated errors of mean cluster redshifts δ_u, are of the same order as the residuals, ρ, with the mean δ_u equal to 0.010 and the mean $|\rho|$ equal to 0.005. This agreement is consistent with the interpretation that the structural redshifts, P, predicted by Eq. (46.2), are expected values of the means and that the residuals, ρ, are attributable to errors of observation. However, of the remaining clusters for which there are only one or two measured redshifts, the $|\rho|$ of 0.006 is lower than expected. If we assume that the velocity dispersion of the remaining clusters is equal to the mean velocity dispersion, 720 km/sec, of the clusters in Table 18, then the mean value of δ_u for these remaining clusters is 0.008.

The fact that the mean δ_u of the second set of clusters, is less than that of the better observed clusters of Table 18 is due in part to the greater distances of the second set. But the low δ_u's may have another explanation. It is observed fact that many clusters are centered on one or two galaxies which are appreciably brighter and larger than other cluster members. It is seen in Table 19 that for nearby well observed clusters, the values of the redshifts of the brightest galaxy, V_B, are very close to the cluster mean $\bar{V}$. If this is true for the more distant clusters, then the observational selectivity of the brightest galaxies in clusters generates a bias which gives better mean redshifts from small samples than is statistically expected.

The effects of contamination of the cluster sample with noncluster members on the value $\bar{V}$ may become large as fainter galaxies are included[23]. This serves to increase the size of the residuals in the well-observed sample. However, no selection criterion has been used for selecting or rejecting individual redshifts to determine the $\bar{V}$'s of Table 17. All published redshifts have been used with equal weight.

Table 18. *Estimated errors for mean cluster redshifts*

Cluster	Name	n	$\bar{V}$	σ_V	$\sigma_V\sqrt{n}$	du/u	δ_u	ρ
....	Virgo	73	1136	643	76	0.067	0.028	−0.0043
0316 + 4121	Perseus	7	5435	715	270	0.049	0.021	0.0006
0123 − 0137	NGC 541	43	5439	450	69	0.012	0.005	0.0006
1257 + 2812	Coma	50	6432	1745	246	0.038	0.016	−0.0035
1627 + 3937	Abell 2199	19	9028	864	198	0.021	0.009	0.0096
1603 + 1755	Hercules	15	10775	652	169	0.015	0.007	−0.0028
2308 + 0720	Pegasus II	3	12821	662	383	0.029	0.012	−0.0097
1145 + 5559	Ursa Major I	4	15269	358	179	0.011	0.005	−0.0097
1520 + 2754	Corona Borealis	8	21651	1294	457	0.020	0.008	0.0042
1534 + 3749		3	45951	408	236	0.005	0.002	−0.0016
0855 + 0321	Hydra	3	60860	137	80	0.001	0.0004	0.0094

Note: Table includes only those clusters with 3 or more observed red shifts.

In order to examine the probability that the agreement between the observed and structural redshifts is not reproducible by chance, we shall use as a statistic the number of fits to within a prescribed tolerance δ occuring between test samples and the discrete values defined by Eq. (46.3). The test samples to be compared are (1) the observed values of the redshifts, and (2) a set of random values drawn from a suitable density distribution defined over the same range as the observed sample, $0.7700 \leqq \log \bar{u}_0 \leqq 2.4300$. If y is the number of clusters with redshift $\bar{u}$, then the density function derived from the envelope of the observed sample is $\log y = 0.76 - 2\log \bar{u}$ over the above range. This function is adopted for the density distribution of the random values.

[23] Godfredsen, E. A.: *Astrophys. J. 139*, 520 (1964).

Table 19. *Comparison of cluster mean redshift with brightest galaxy redshift*

Cluster	Name	Brightest galaxy	V_B	$\lvert\bar{V}-V_B\rvert$
....	Virgo	NGC 4486	1171	35
0316 + 4121	Perseus	NGC 1275	5293	142
0123 − 0137	NGC 541	NGC 547	5472	33
0257 + 2812	Coma	NGC 4889	6428	4
1627 + 3937	Abell 2199	NGC 6166	9082	54

The results of two Monte Carlo experiments programed for the RAND 7044 computer are given in Table 20. In Table 20a, the observed sample of 28 redshifts is compared with 25 sets of 28 random values

Table 20. *Comparison of observational and random fits*

δ	N(obs)	$\bar{N}_R$	$N_{R\max}$	σ_R	t	Probability
(*a*) 0.020	26	24.56	27	1.24	1.17	0.242
0.010	19	16.40	18	0.98	2.65	0.008
0.005	14	11.28	14	1.04	2.62	0.009
0.003	10	8.36	11	1.05	1.56	0.119
(*b*) 0.020	28	27.44	28	0.50	1.13	0.258
0.010	24	20.44	23	1.13	3.14	0.002
0.005	17	14.48	16	0.98	2.56	0.010
0.003	12	10.80	12	0.85	1.41	0.159

selected from the above density distribution. The comparison in Table 20 is between observational and random data fits to Eq. (46.1) written as

$$P = A - \log N(N+1),$$

where A is independently selected for the observed redshifts and for each of the 25 sets of 28 random redshift values. In the observational and in each random case, the machine program selects the value of A which gives the maximum number of fits within the prescribed δ. For the observed redshifts, the machine-selected A equals 3.4976. (This is the same as $\sqcup - \log M(M+1)$ with M set equal to 2 and $\sqcup = 4.2758$.)

The first column in Table 20 gives the prescribed δ's defining the allowed margins of fit, $P \pm \delta$. The second column, N(obs), gives the number of fits (within δ) to the 28 observed redshifts, using the above equations with $A = 3.4976$. N_R is the number of fits (within δ) occuring in any one sample of 28 random redshift values. There are 25 sets of computer-generated N_R's. The third column of Table 20 lists $\bar{N}_R$, the mean of the values of N_R occurring in all 25 sets for a prescribed δ. $N_{R\max}$ is the largest number of fits in the 25 sets, and σ_R is the dispersion in N_R.

In the sixth column, t is the standardized variable, $[N(\text{obs}) - \bar{N}_R]/\sigma_R$. The last column gives an estimated probability, derived from the standard error integral, assuming a normal distribution for N_R.

It is interesting that the largest value of t occurs at the value of δ which equals the estimated observational error, δ_u. This is expected if the values of N_R derive from a unimodal or monatonic density distribution and the values of $N(\text{obs})$ derive from a multimodal density distribution consisting of a set of error functions whose means are located at the discrete values $P(2, N)$ and whose standard deviations are equal to δ_u. In this situation, the size of N_R will decrease uniformly with decreasing δ, while the size of $N(\text{obs})$ will be relatively insensitive to δ until $\delta = \delta_u$. For $\delta < \delta_u$, $N(\text{obs})$ will decrease sharply. This results in the difference, $N_R - N_{(\text{obs})}$, having a maximum near $\delta = \delta_u$.

Table 20b is the same as 20a except thet the comparison of the observed redshifts and random redshifts is for the $M = 2$ and $M = 4$ sequences taken together. Tables 20a and 20b both show that for the significant value of $\delta = \delta_u = 0.010$, the probability is minimum.

The test of statistical significance is summarized in Table 21. We recall that Eqs. (46.1) and (46.2) were originally derived from relations between mean redshifts and diameters in five clusters (Coma, Ursa Major I, Corona Borealis, Boötes, and Ursa Major II). The original relation based on these five clusters was,

$$\log \bar{u} = 4.2792 - \log N(N+1) - \log M(M+1).$$

If this relation is adopted as the *complete* hypothesis, including the value of the constant, $\sqcup$, and is tested against the remaining 23 clusters, we get the results reported in Table 21. Here we compare the number of fits in the *remaining* 23 clusters to the values based on the original relation with the number of fits occurring in 25 sets of 23 random redshift values. In this test the value of the constant in each set of random values is not floating to maximize the number of fits, but is held *fixed* at the predicted value, $\sqcup = 4.2792$. The most significant level of fit again occurs at $\delta = \delta_u$ with the probability of the number of observed fits being accounted for by a chance mechanism being 1 in 1,000.

Table 21. *Comparison of observational and random fits to Eq. (46.2)*

δ	$\bar{N}(\text{obs})$	$\bar{N}_R$	$N_{R\max}$	σ_R	t	Probability
0.020	23	18.40	21	2.32	1.99	0.047
0.010	20	11.76	16	2.45	3.36	0.001
0.005	13	6.92	11	2.30	2.65	0.008

A conservative assessment of the statistical results of Tables 20 and 21 allows the hypothesis of regularized structure in the distribution of clusters, as expressed by Eq. (46.2), at least equal admissibility with the hypothesis of random distribution. The value of t actually suggest that greater weight be given to the hypothesis of discretized structure. However, in the absence of better theoretical justification for Eq. (46.2), it cannot be assumed that it is the best representation for the discretized structure which seems to exist.

47. Other Redshift Discretizations

The statistical tests of Section 46 show that there exists at least one discretized representation for the mean synoptic redshifts of clusters of galaxies (namely, that of Eq. (46.2)) that has a less than one part in a thousand probability of having originated by chance. That this particular representation best describes whatever regularity exists in the distribution of the redshifts cannot be claimed, but the existence of one such distribution implies that the distribution of the mean redshifts of clusters of galaxies is not completely random. The salient point of this result is the inference for the existence of regularized structure in the large-scale distribution of matter. If we may assume for the present that Eq. (46.2) represents the cosmic distribution of clusters, it follows that the allowable values for the temporal and radial coordinates of matter concentrations (as seen by an observer located at one of these concentrations) are those that would correspond to clusters being located on a set of concentric shells that possess a definite relation between successive radii, with the permissible spatial locations being at the intersections of the sets of cluster-centered concentric shells.

This predicted structure in the *angular* distribution of cluster centers as seen by equivalent observers indicates one way in which the investigation of regular structure in the distribution of clusters may be investigated further. By analyzing the angular distributions of the cluster for regularities or by combining the angular positions of clusters with their mean redshifts to generate additional "quasi-redshifts" by triangulation, the present small sample of cluster redshifts may be effectively enlarged.

The measurement of redshifts of galaxies beyond the local supercluster has been primarily to determine the law of redshifts and secondarily to study the dynamics of rich regular clusters such as Coma and Abell 194. The selectivity affects of these two programs have resulted in the bulk of measurements being distributed among a very small sample of less than 30 clusters. Redshifts of galaxies in even the nearby clusters can be observed in only the largest telescopes, so it may be a good many years before a satisfactory sample of mean cluster redshifts may be amassed. An alternative way to test the hypothesis of regularized distribution of clusters is accordingly welcome. In using angular position data of cluster centers to generate "quasi-redshifts", we assume redshifts to be proportional to distance and space to be approximately Euclidean for z's up to 0.05. We may then use the law of cosines for plane triangles to determine the distance between two clusters whose sky positions and redshifts are known. In this way the values of the redshifts of clusters as measured by observers located in other clusters may be derived.

The sample of observed cluster redshifts given in Table 17 has been augmented with redshifts of radio sources[24] which in many cases are identifiable with giant ellipticals and in the cases which may be checked, have redshifts near the value of the cluster mean. The augmented sample of 45 objects – clusters and radio sources – has been used to generate quasi-redshifts, which (as far the assumptions of proportionality and flatness of space are valid) represent the distances between the 45 clusters.

If the clusters are distributed at random, the histogram of quasi-redshifts will present a noise-like spectrum. At the other extreme if there exists a *single* regular structure, the spectrum should consist of a set of sharp resonances or peaks. Actually, the spectrum is found to consist of several singular peaks which emerge from a dense set of lower peaks in the manner of signal and noise. Only those quasi-redshifts falling in the range of redshifts given in Table 17 ($z < 0.20$) were retained in the analysis.

As an example of the statistical analysis used to test the significance of resonances, the statistics of the resonance at 10,280 km/sec are given. In Table 22 are listed the pairs of objects whose separation correspond through Hubble's Law to a redshift or quasi-redshift with a value near 10,280 km/sec. The redshift of ML 518 is the only directly observed redshift, the others are derived from observed redshifts and angular separations. The midvalue, M, is 10281 km/sec; the spread, S, is 66 km/sec. Half the spread is near the value of the mean error of an individual observation. A range, r, of 2000 km/sec (centered on M) was selected to assure that the value of the random density function would be the *local* mean density. The total population in this range is $n = 59$ redshifts. The probability of a random redshift falling in the resonance interval $M \pm S/2$ is $p = S/r = 0.033$. Determining the probability of k successes in n trials in the usual way for binominal distributions, with $np = 1.95$; $\sigma = 1.38$; $k = 7$; and the resulting score $(k - np)/\sigma = 3.68$, corresponds

[24] Maltby, P., Mathews, J. A., Moffet, A. T.: *Astrophys. J. 137,* 153 (1963); Schmidt, M.: *Astrophys. J. 141,* 1 (1965).

to a probability of 2 parts in 10^4 that the observed number of successes can be obtained by chance from a uniform probability density over the range r.

Table 22. *Example of redshift/quasi-redshift resonance*

Object paris	Pair separation (km/sec)
A 1656 – 3 C 66	10314
3 C 442 – 3 C 353	10300
A 426 – A 1213	10271
A 1377 – A 1656	10269
Local – ML 518	10253[a]
3 C 270 – 3 C 78	10252
3 C 278 – 3 C 66	10248

[a] Observed redshift of ML 518.

Table 23 summarizes the statistics for the ten resonances in the range 5000 to 45000 km/sec. M is the midvalue, S the spread, k the number of redshifts in the resonance, x the expected number of successes, and $\frac{k-x}{\sigma}$ the measure of statistical significance with corresponding probabilities given in the right column. All resonances with scores of $\frac{k-x}{\sigma} > 3$ (probability 3 parts in 10^3) will be taken as unlikely random fluctuations. The resonance at 5461 km/sec agrees with the *observed* mean redshifts of 5435 km/sec (from 7 galaxies) of the Perseus cluster and 5439 km/sec (from 43 galaxies) of the A 194 cluster. Five quasi-redshifts also have this value. It might be suspected that peaks in quasi-redshifts are results of proliferation of coincidences of two or more observed redshifts, but a check of the objects associated with each peak shows this factor is not a contribution.

Table 23. *Resonances in quasi-redshifts to 45,000 km/sec*

M	k	S	x	$\frac{k-x}{\sigma}$	P
5461	7	85	1.41	4.72	<0.0001
9018	6	48	1.42	3.88	0.0001
10281	7	66	1.95	3.68	0.0002
11535	6	68	1.49	3.73	0.0002
14677	6	85	1.98	2.86	0.004
18672	9	152	3.48	3.00	0.003
21566	10	175	2.61	4.62	<0.0001
25343	5	83	1.24	3.38	0.0007
42519	8	102	1.89	4.56	<0.0001
43152	5	86	1.59	2.78	0.005

In addition to it being highly unlikely that the resonances are attributable to random fluctuations, the mean values of the resonances are not randomly distributed but bear simple ratios to one another. The mean values of the resonances out to 19,000 km/sec are given by the scale factor 1720 km/sec times the numbers of the set: $\sqrt{10}$, $3\sqrt{3}$, $2\sqrt{7}$, 6, $3\sqrt{5}$, $6\sqrt{2}$, $3\sqrt{13}$ with residuals of the order of 50 km/sec[25]. Beyond 20,000 km/sec the fits to simple ratios disappear. The interesting thing about these numbers is they are the ratios of the distances between centers of closely packed uniform spheres arranged in a cuboctahedral pattern[26]. This suggests that a histogram of the observed angles of separation of the clusters might possess peaks at the angles expected for closely packed spheres. The large numbers of clusters observed in all unobscured directions in the sky renders statistically meaningless any patterns selected *ab initio* on the basis of angular distribution criteria, accordingly, an independent selectivity factor must be invoked. A study was made of the clusters in Abell's catalogue[27] selected on the basis of membership in the richest classes (4 and 5). Though widely separated, these clusters have angular positions consistent with structured rather than random distribution. In addition, statistically significant fits to some cuboctahedral angles are found for a sample of clusters selected on the basis of having measured redshifts. The number of fits to $\pm$ 35 km/sec between quasi-redshifts for a given cluster and the cuboctahedron ratio sequence is found to depend on the richness class of the cluster. The percent of successes varies as follows: 0.34 for radio sources (not identified with clusters), 0.36 for Abell richness class 0, 0.47 for richness class 1, and 0.55 for richness class 2.

These representations suggesting non-random structure on a scale up to 10^9, parsecs pose some fundamental cosmogonic questions. In view of the same difficulties which arise in explaining second order clusters as dynamic systems[28], it is unsupportable to postulate the existence of a dynamic system with a diameter of the order of 10^9 parsecs, the size consistent with the distances and angular separation of the clusters involved. Consequently, if the apparent superorganizations to which these clusters belong are real, they must either originate through some physical communication process other than those presently recognized or must be the vestige of a structure which existed in an early stage of

---

[25] Wilson, A. G.: *Astron. J. 70*, 150 (1965).

[26] Uniform spheres packed in layers according to the arrangement ABCABCABC ... have centers on the surfaces of regular cuboctrahedra. With layer A as the reference layer, B layer is defined as shifted "down" by $2/\sqrt{3}$ sphere radii and C layer is defined as shifted "up" by $2/\sqrt{3}$ sphere radii.

[27] Abell, G. O.: *Astrophys. J. Suppl.* No. *31*, 3, 211 (1958).

[28] Zwicky, F.: *Pub. Astron. Soc. Pacific 69*, 518 (1957).

the universe when matter was highly compact. In a universe that expanded from a "primeval atom" according to evolutionary models, the structure observed at present may be the vestige of a highly compact regular "crystal like" structure that existed during a brief time in the earlist stages of the expansion. In the case of an oscillating model with the universe "bouncing off" of the Schwarzschild singularity, the period of high compaction was of sufficient duration to force concentrations of matter corresponding to clusters of galaxies into a more or less close packed configuration that preserved the relative separations and angular features of its structure under the subsequent uniform dialation.

The existence of a structured distribution of cosmic bodies would provide a consistency test for the "vestige hypothesis". If some regular structure, such as a cuboctahedral structure, resulted from physical action that took place during the highly compacted stage of the universe, and this structure was preserved during a subsequent isotropic expansion, then the "isochronous world map" should possess a better fit to the structure than the "world picture", i. e., the resonances and their ratios corrected for light travel time should have greater statistical significance than those based on uncorrected parameters. The discrepancy between the map and picture also should be a function of the mean value of the resonance.

Comparisons of resonances in the cluster sample on which the results of this section are based show no significant differences between map and picture. However, the detection by Burbidge, Cowan *et al.*[29] of regular discretized patterns in the redshifts of quasars where values of z are up to 10 times larger than in the cluster sample might reveal significant differences.

In studying the distribution of emission line redshifts, Burbidge reports a uniform discretization sequence of the form $z = (0.061)n$, with a marked peak at $n = 1$ containing 5 objects in different parts of the sky with redshifts of either 0.060 or 0.061[30]. This peak closely corresponds to the quasi-redshift resonance in Table. 23 at 18,670 km/sec consisting of 9 quasi-redshifts (but no observed redshifts)[31]. Although Burbidge's other peaks lie outside the range of the resonances in Table 23, further comparisons between the distributions of the cluster redshifts

---

[29] Burbidge, G.: *Astrophys. J. 154,* L 41 (1968); Cowan, C. L.: *Astrophys. J. 154,* L 5 (1968); Barbieri, C., Bonometto, S., Saggion, A.: *Astrophysical Letters 2,* 225 (1968); Burbidge, G. R., Burbidge, E. M.: *Nature 222,* 735 (1969).

[30] Burbidge, G.: *Astrophys. J. 154,* L 41 (1968).

[31] The generation of quasi-redshifts to augment the sample of observed redshifts reported here was carried out in 1964 before the observation of the redshifts reported by Burbidge. The prediction of the 0.061 resonance supports the usefulness of the quasi-redshift strategy.

and the emission redshifts could be made on the basis of a set of quasi-redshifts based on the emission values. This may prove to be an important clue in identifying the nature of quasar redshifts. A high degree of correlation between quasar and cluster quasi-redshift peaks would reinforce the interpretation of quasar redshifts as being largely attributable to the same cause as the cluster redshifts, i.e., cosmic. Little or no correlation on the other hand would support the hypothesis of quasar redshifts being due to other causes.

48. Atomic-Cosmic Relations

In addition to the many physical theories that relate atomic and astrophysical processes, it has long been known that the dimensionless *numbers* of microphysics, such as the ratio of electric to gravitational forces, $S = e^2/Gm_p m_e$, frequently put in an appearance in macrophysics; as for example, the occurence of S^2 in equations of cosmic mass[32] or the occurence of $(hc/G)^{\frac{1}{2}}$ in expressions for stellar mass[33] or the fine structure constant, $\alpha = 2\pi e^2/hc$, appearing in the ratio of the potential limit of stable, non-degenerate cosmic bodies to the Schwarzschild relativistic potential limit[34]. It is not known whether the appearance of these basic microphysical numbers in astronomical measurements is due to their being reflected by atomic processes in astrophysics or is due to these numbers being fundamental to physical processes on whatever level or scale. This subject is introduced here because some of the bases resulting from compaction of the functions of astronomical observables in Sections 43, 44, and 46 also appear to reflect microphysical values.

Redshifts and synoptic redshifts are dimensionless observables and if the results of Section 46 are valid, the compaction base ⧢ given in Eq. (46.2) should be an important dimensionless constant of cosmic physics. The compaction base for 17 of the clusters is 3.4977 and for 6 of the clusters is 2.9723. These two bases compact for $M = 2$ and $M = 4$ onto 4.2759 and 4.2733 respectively. The weighted mean is 4.2752. This is the common logarithm of $(137.28)^2$. If the "coincidence" with the value $1/\alpha = 137.0388$[35] (which lies within the probable error limits) may be assumed a non-chance coincidence, Eq. (46.2) can be written

$$\bar{u}_{M,N} = \alpha^2 M(M+1)N(N+1). \tag{48.1}$$

[32] For a summary discussion of these cosmic numbers see Bondi, H.: *Cosmology*. Cambridge: Cambridge University Press 1952.

[33] Chandrasekhar, S.: *Nature 139*, 757 (1937).

[34] Wilson, A. G.: *Science 165*, 202 (1969); Wilson, A.: Hierarchical Structure in the Cosmos. In: *Hierarchical Structures*. Eds. Whyte, L. L., Wilson, A., and Wilson, D. New York: American Elsevier 1969.

[35] Cohen, E. R., DuMond, J. W. N.: *Phys. Rev. 37*, 537 (1965).

That is to say that the observed mean synoptic redshifts of clusters of galaxies may be mapped onto a discrete two parameter grid, each of whose scale factors is equal to the structural constant α.

If the *diameter* equations are expressed in dimensionless from, then their compaction bases should also represent dimensionless constants of cosmic importance. Whenever the synoptic redshift is used, the only correction required to express the equations in dimensionless form is the scale factor converting the angular diameters to radians.

For the bright galaxies from de Vaucouleurs' Catalogue[36], D, the angular diameter is given in units of 0.1 min of arc. If θ is in radians, then

$$\log\theta = \log D + \log[\pi/(180 \times 60 \times 10)] = \log D - 4.5363\,.$$

Since the compaction algorithm used in Section 43 is independent of additive constants, the bright galaxy compaction equation,

$$\{\{1 + \log D - \tfrac{4}{9}\log u\} \dot{-} E_2(n)\} \dot{+} E_3(v) = 1.043\,,$$

may be written in the dimensionless form,

$$\{\{\log\theta - \tfrac{4}{9}\log u\} \dot{-} E_2(n)\} \dot{+} E_3(v) = 1.043 - 1 - 4.536 = -4.493\,. \quad (48.2)$$

For the cluster galaxies discussed in Tables 12 and 13, the scale factor is $\log\theta - \log D = -4.8658$. Substituting in Eq. (44.2) gives,

$$\{\{\log\theta - \tfrac{2}{3}\log\bar{u}\} \dot{-} E_2(n)\} \dot{+} E_3(v) = 1.069 - 4.866 = -3.797\,. \quad (48.3)$$

Comparisons of the dimensionless compaction bases in Eqs. (48.1), (48.2), and (48.3) show that all are of the same order of magnitude. In the one case of the three in which both the systematic errors and internal errors are known to be small, the value of the base is very close to $\log\alpha^2 = -4.2737$. In the other two cases, while the discrepancy between the values of the bases and $\log\alpha^2$ greatly exceeds the internal error, the unknown systematic errors in diameter measurements and the problem of translating operationally defined diameters into physically consistent diameters precludes any significance being attached to the exact values of the dimensionless base. While the uncertainties in the diameter measurements have been such that *relative* values have been consistent and permit meaningful comparisons within a set to be made, no claims can be made for the absolute values. It would be quite disturbing, however, to find wide discrepancies in the values of the dimensionless compaction bases. Their clustering to within a half order of magnitude well meets the "customary astronomical standard" for absolute quantities.

---

[36] Vaucouleurs, A. and G. de: *Reference Catalogue of Bright Galaxies*. Austin: University of Texas Press 1964.

Appendix

A. Curvature Conventions and Signs

There are two basic sources of confusion in general relativity, and both of them come from different conventions used in the literature. The first arises from the fact that the metric tensor of an Einstein-Riemann space may have either signature (-2) or signature $(+2)$; that is, the metric tensor at a point may be reduced to either diag$(1, -1, -1, -1)$ or diag $(-1, 1, 1, 1)$. The convention adopted here is that the signature is (-2). The reader is referred to Synge[1] where a full table of conversions is given between the two choices of signature.

The second source of confusion arises because there are several ways of defining the curvature tensor of a metric space. We follow the notation and conventions of Schouten[2] and hence we have

$$R^{D}_{ABC} = 2\partial_{[A}\Gamma^{D}_{B]C} + 2\Gamma^{D}_{[A|E|}\Gamma^{E}_{B]C}, \tag{A-1}$$

$$\nabla_{[A}\nabla_{B]}V^{D} = \tfrac{1}{2}R^{D}_{ABC}V^{C}, \tag{A-2}$$

and

$$\nabla_{[A}\nabla_{B]}W_{C} = -\tfrac{1}{2}R^{D}_{ABC}W_{D}. \tag{A-3}$$

Although the symbols on the right-hand side of (A-1) remain the same, numerous other orderings of the indices on the left-hand side of (A-1) appear. For instance, if we use $\{K^{D}_{ABC}\}$ to denote the form used by Synge, we have

$$K^{D}_{BCA} = R^{D}_{ABC}, \tag{A-4}$$

and hence, since $R_{AB} = R^{D}_{DAB}$ and $R^{D}_{(AB)C} = 0$,

$$R_{AB} = -K_{AB} \tag{A-5}$$

and the Einstein tensors of Synge and of Schouten differ in sign. Accordingly, with Schouten's notation, the Einstein equations read

$$G_{AB} = R_{AB} - \tfrac{1}{2}Rh_{AB} = \kappa T_{AB}, \tag{A-6}$$

while for Synge's notation we have

$$G_{AB} = K_{AB} - \tfrac{1}{2}Kh_{AB} = -\kappa T_{AB}. \tag{A-7}$$

---

[1] Synge, J. L.: *Relativity: The General Theory*. Amsterdam: North-Holland 1960.

[2] Schouten: p. 238ff.

Appendix

B. The Spherically Symmetric Case

An Einstein-Riemann space is spherically symmetric if and only if a coordinate system can be found such that the fundamental differential form is given by

$$ds^2 = V(x^0,x^1)^2(dx^0)^2 - F(x^0,x^1)^2(dx^1)^2 - G(x^0,x^1)^2\,dL^2 \tag{B-1}$$

where

$$dL^2 = (dx^2)^2 + \sin^2 x^2 (dx^3)^2 \tag{B-2}$$

is the metric on the surface of the unit sphere in three-dimensional Euclidean space[3]. In such spaces, a spherically symmetric bounding surface $\mathscr{S}$ is defined by the equations

$$f(x^0,x^1) = x^1 - y(x^0) = 0 \tag{B-3}$$

and hence a parametric representation is given by

$$x^0 = u^0\,, \quad x^1 = y(u^0)\,, \quad x^2 = u^1\,, \quad x^3 = u^2\,. \tag{B-4}$$

A simple calculation then gives the following values for the components of the first fundamental form:

$$a_{00} = \bar{V}^2 - (y')^2 \bar{F}^2\,, \quad a_{11} = -\bar{G}^2\,, \quad a_{22} = -\bar{G}^2 \sin^2 u^1\,,$$
$$a_{\Gamma\Sigma} = 0 \quad \text{for} \quad \Gamma \neq \Sigma\,. \tag{B-5}$$

The superimposed bar is used to denote that the corresponding quantity is obtained under the substitution (B-4).

The assumption of spherical symmetry places severe restrictions on the form of the momentum-energy tensor. It can be shown[4] that the assumption of spherical symmetry is consistent with the Einstein field equations if and only if, for a coordinate system for which (B-1) holds, we have

$$T_{22}h^{22} = T_{33}h^{33}\,, \; T_{12} = T_{23} = T_{31} = 0\,, \; \partial_2 T_A^B = \partial_3 T_A^B = 0\,. \tag{B-6}$$

Since these conditions must hold on both sides of $\mathscr{S}$, the jump strengths of the momentum-energy tensor assume the form

$$((S_A^B)) = \begin{pmatrix} a & b & 0 & 0 \\ b' & c & 0 & 0 \\ 0 & 0 & d & 0 \\ 0 & 0 & 0 & d \end{pmatrix}, \quad b h_{11} = b' h_{00}\,, \tag{B-7}$$

---

[3] Takeno, H.: *The Theory of Spherically Symmetric Space-Times*. Scientific Reports of the Research Institute for Theoretical Physics, Hiroshima University, No. 3 (1963).

[4] Synge, J. L.: *Relativity: The General Theory*. Amsterdam: North-Holland 1960.

where a, b, c and d are functions of x^0 and x^1 only. From this we see that the surface tensor $\{S^\Sigma_\Gamma\}$ has the form

$$((S^\Sigma_\Gamma)) = \begin{pmatrix} A & 0 & 0 \\ 0 & d & 0 \\ 0 & 0 & d \end{pmatrix}, \qquad A = a^{00}(a\bar{h}_{00} + 2b\bar{h}_{11}y' + c\bar{h}_{11}y'^2). \tag{B-8}$$

Each component of $\{S^\Sigma_\Gamma\}$ is thus a function of u^0 only. We also note that we may write $\{S_{\Gamma\Sigma}\}$ as

$$S_{\Gamma\Sigma} = L^{-2} a_{00}(A-d)L_\Gamma L_\Sigma + d\,\delta^\Sigma_\Gamma \tag{B-9}$$

where

$$L_\Gamma = L\delta^0_\Gamma \tag{B-10}$$

defines the components of a surface vector.

If we assume that the three-dimensional space $\mathscr{S}$ with fundamental form $a_{\Gamma\Sigma}\mathrm{d}u^\Gamma \mathrm{d}u^\Sigma$ is *static*, the fact that $\partial_1 a_{\Gamma\Sigma} = 0 = \partial_2 a_{\Gamma\Sigma}$ implies that the function $y(u^0)$ must be such that the functions $V(u^0, y(u^0)) - F(u^0, y(u^0))(y')^2$ and $G(u^0, y(u^0))$ must be independent of u^0. The surface $\mathscr{S}(t)$ thus moves in $\mathscr{R}(t)$ is such a fashion that it maintains the same value of $G(u^0, y(u^0))$; that is, $\mathscr{S}$ is in geometric equilibrium with the Einstein field. In addition, the functions $S_{\Gamma\Sigma}$ are u^0 independent and the vector $\{L_\Gamma\}$ is parallel to the vector $\{U_\Gamma\}$ of Chapter II. If it is also assumed that $\mathscr{E}$ is stationary in a neighborhood of $\mathscr{S}$, the quantities A, a, b, c, d, a_{00} and $\{\bar{h}_{AB}\}$ are independent of u^0, in which case $\{S_{\Gamma\Sigma}\}$ is a constant valued tensor field.

Appendix

C. On Conformally Related Metric Spaces, One of which Admits an Irrotational Isometry

1. The Space $\mathscr{H}_n$

Let $\mathscr{H}_n$ denote an n-dimensional ($n > 2$) hyperbolic-normal metric space with fundamental differential form

$$\mathrm{d}s^2 = h_{AB}\mathrm{d}x^A \mathrm{d}x^B, \tag{C 1.1}$$

the hyperbolic-normal condition being reflected by the fact that the signature of $((h_{AB}))$ is $2-n$. As is customary, a vector field on $\mathscr{H}_n$, with components V^A, is referred to as timelike, spacelike, or null accordingly as $V^A V^B h_{AB}$ is greater than, less than, or equal to zero.

Assumption 1.1 *The structure of the space $\mathscr{H}_n$ is such that it admits a time-oriented irrotational isometry.*

The particular properties of $\mathscr{H}_n$ that are salient to our considerations are delineated in the following lemma; the reader is referred to any one of a number of standard texts for the proof.

Lemma 1.1 *Under Assumption 1.1, there exists a vector field* $\{Y^A\}$ *and a scalar field* ψ *on* $\mathscr{H}_n$ *such that*

$$Y_A Y^A = \exp(-2\psi), \quad U_A = Y_A \exp(\psi), \tag{C 1.2}$$

$$\nabla_B U_A = U_B \partial_A \psi, \quad U^A \partial_A \psi = 0, \quad U^B \nabla_B U_A = \dot{U}_A = \partial_A \psi, \tag{C 1.3}$$

$$K_{AB} U^B = (h^{CD} \nabla_C \nabla_D \psi) U_A, \tag{C 1.4}$$

where $\{K_{AB}\}$ *is the Ricci tensor of* $\mathscr{H}_n$.

Definition 1.1. The vector field $\{U^A\}$ and the scalar field ψ are referred to as the *basic* vector and scalar fields of the time-oriented irrotational isometry of $\mathscr{H}_n$.

2. The Conformally Related Space $\mathscr{C}_n$

Let $\mathscr{C}_n$ denote an n-dimensional, hyperbolic-normal metric space with fundamental differential form

$$d\bar{s}^2 = \bar{h}_{AB}\, dx^A\, dx^B. \tag{C 2.1}$$

We assume that $\mathscr{C}_n$ and $\mathscr{H}_n$ have the same coordinate patches and coordinate functions (i.e., $\{dx^A\}$ in $\mathscr{H}_n$ is the same as $\{dx^A\}$ in $\mathscr{C}_n$, and the respective points of $\mathscr{H}_n$ and $\mathscr{C}_n$ have the same coordinates with respect to the coordinate systems in which $\{dx^A\}$ are computed).

Definition 2.1. If there exists a scalar function λ on $\mathscr{H}_n$ (on $\mathscr{C}_n$) such that

$$h_{AB} = \bar{h}_{AB} \exp(2\lambda), \quad (h^{AB} = \bar{h}^{AB} \exp(-2\lambda)), \tag{C 2.2}$$

holds at all points of $\mathscr{H}_n$ (at all points of $\mathscr{C}_n$), the space $\mathscr{C}_n$ is said to be conformally related to the space $\mathscr{H}_n$ with parameter λ. We assume throughout this discussion that the conditions of Definition 2.1 are met. It must be clearly noted that both $\mathscr{H}_n$ and $\mathscr{C}_n$ are well-defined metric spaces, in contrast to the usual case in which $\mathscr{C}_n$ is considered as a conformal space with no metric structure of its own.

The separate metric structures of $\mathscr{C}_n$ and $\mathscr{H}_n$ require certain notational conventions for adequate distinction between similar quantities and operations in the two spaces.

Agreement 2.1. Basic geometric quantities will be written in the usual fashion when they refer to $\mathscr{H}_n$. If these quantities are formed in $\mathscr{C}_n$, they will be written with the same kernel letter but with a bar superimposed above.

Thus, the Ricci tensor of $\mathscr{H}_n$ is written as

$$K_{AB} = \partial_C \Gamma^C_{AB} - \partial_B \Gamma^C_{CA} + \Gamma^C_{CD} \Gamma^D_{AB} - \Gamma^C_{DA} \Gamma^D_{CB}, \tag{C 2.3}$$

while in $\mathscr{C}_n$ we write $\{\bar{K}_{AB}\}$. This convention has already been observed in writting (C 1.1) and (C 2.1).

Agreement 2.2. The metric tensor in $\mathscr{H}_n$ will be used to raise and lower indices in the usual fashion. If the metric tensor of $\mathscr{C}_n$ is used for this purpose, a dot will be placed in the original position of the index.

Thus, if V_A is a covariant vector field, we have

$$V^A = h^{AB} V_B, \qquad V^{.A} = \bar{h}^{AB} V_B.$$

Agreement 2.3. Covariant differentiation formed from the Christoffel symbols of H_n will be denoted by ∇, while $\bar{\nabla}$ will be used to denote covariant differentiation formed on the Christoffel symbols of $\mathscr{C}_n$.

Note that certain well-known results from the theory of conformal spaces carry over directly to conformally related metric spaces under suitable interpretations. The method of proof will only be indicated, since the details can be found in any adequate text on differential geometry.

As an immediate consequence of the definition of the Christoffel symbols of the second kind, we have

$$\Gamma^A_{BC} = \bar{\Gamma}^A_{BC} + P^A_{BC}, \tag{C 2.4}$$

where the quantities P^A_{BC} constitute the components of a tensor (of indicated type):

$$P^A_{BC} = \delta^A_B \partial_C \lambda + \delta^A_C \partial_B \lambda - \bar{h}_{BC} \partial^{.A} \lambda. \tag{C 2.5}$$

When we substitute (C 2.5) into (C 2.3) and use the definition of the covariant derivative, we obtain

$$K_{AB} = \bar{K}_{AB} + \bar{\nabla}_C P^C_{AB} - \bar{\nabla}_B P^C_{CA} + P^C_{CD} P^D_{AB} - P^C_{DA} P^D_{CB}. \tag{C 2.6}$$

Thus, since (C 2.5) leads to

$$\begin{aligned}
\bar{\nabla}_B P^C_{CA} &= n \bar{\nabla}_B \partial_A \lambda, \qquad \bar{\nabla}_C P^C_{AB} = 2 \bar{\nabla}_B \partial_A \lambda - \bar{h}_{AB} \bar{\nabla}_C \partial^{.C} \lambda, \\
P^C_{CD} P^D_{AB} &= n (2 \partial_A \lambda \partial_B \lambda - \bar{h}_{AB} \partial_C \lambda \partial^{.C} \lambda), \\
P^C_{DB} P^D_{CA} &= (n+2) \partial_A \lambda \partial_B \lambda - 2 \bar{h}_{AB} \partial_C \lambda \partial^{.C} \lambda,
\end{aligned}$$

the system (C 2.6) gives the following explicit evaluations:

$$\begin{aligned}
K_{AB} = \bar{K}_{AB} &- (n-2) \bar{\nabla}_B \partial_A \lambda + (n-2) \partial_A \lambda \partial_B \lambda \\
&- \bar{h}_{AB} (\bar{\nabla}_C \partial^{.C} \lambda + (n-2) \partial_C \lambda \partial^{.C} \lambda),
\end{aligned} \tag{C 2.7}$$

$$K = \exp(-2\lambda) \{\bar{K} - 2(n-1) \bar{\nabla}_A \partial^{.A} \lambda - (n-1)(n-2) \partial_A \lambda \partial^{.A} \lambda\}, \tag{C 2.8}$$

$$\begin{aligned}
G_{AB} = \bar{G}_{AB} &- (n-2) \bar{\nabla}_B \partial_A \lambda + (n-2) \partial_A \lambda \partial_B \lambda \\
&+ \bar{h}_{AB} \{(n-2) \bar{\nabla}_C \partial^{.C} \lambda + (n-2)(n-3) \partial_C \lambda \partial^{.C} \lambda\}.
\end{aligned} \tag{C 2.9}$$

The quantities G_{AB} constitute the components of the Einstein tensor $\{G_{AB}\}$.

3. Lemma and Relations

The results stated by (C 2.7) allow us to represent the Ricci tensor of $\mathscr{H}_n$ in terms of the Ricci tensor of $\mathscr{C}_n$ together with λ and its derivatives. In order to take advantage of this situation as a means of examining the implications of the time-oriented irrotational isometry in $\mathscr{H}_n$, we shall need the following lemmas.

Lemma 3.1 *We have*

$$h^{AB} \nabla_A \nabla_B \psi = \exp(-2\lambda)(\bar{\nabla}_A + (n-2)\,\partial_A \lambda)(\partial^{A}_{\cdot} \psi). \qquad \text{(C 3.1)}$$

Proof. From the definition of the covariant derivative and (C 2.4), we obtain

$$\nabla_A \nabla_B \psi = \nabla_A \partial_B \psi - P^C_{AB}\, \partial_C \psi .$$

Hence, substituting (C 2.5) into the above, multiplying by $\{h^{AB}\}$ and making use of (C 2.2), we obtain (C 3.1).

Lemma 3.2 *We have*

$$U_{\dot{A}} = \exp(-2\lambda)\, U_A, \qquad U^A U_{\dot{A}} = \exp(-2\lambda). \qquad \text{(C 3.2)}$$

Proof. From the definition of lowering an index and (C 2.2), we have

$$U_A = h_{AB}\, U^B = \exp(2\lambda) \bar{h}_{AB}\, U^B = \exp(2\lambda)\, U_{\dot{A}},$$

and hence the first of (C 3.2) is established. One may easily verify from (C 1.2) that $\{U_A\}$ is a unit timelike vector field on $\mathscr{H}_n$. We thus have

$$1 = U^A U_A = U^A U_{\dot{A}} \exp(2\lambda),$$

and hence the second of (C 3.2) is established.

Lemma 3.3 *The covariant derivative of the vector field $\{U^A\}$ assumes the following form when calculated in $\mathscr{C}_n$:*

$$\bar{\nabla}_B \mathrm{U}^A = \exp(-2\lambda)\, \mathrm{U}_B\, \partial^{A}_{\cdot} (\psi + \lambda) - \delta^A_B\, \mathrm{U}^C\, \partial_C \lambda - U^A\, \partial_B \lambda . \qquad \text{(C 3.3)}$$

Proof. Making use of (C 1.3), (C 2.2), (C 2.4) and (C 2.5), the following calculation results:

$$\begin{aligned} \nabla_B U^A &= U_B\, \partial^A \psi = \exp(-2\lambda)\, U_B\, \partial^{A}_{\cdot} \psi = \bar{\nabla}_B U^A + P^A_{BC}\, U^C \\ &= \bar{\nabla}_B U^A + \delta^A_B\, U^C\, \partial_C \lambda + U^A\, \partial_B \lambda + \bar{h}_{BC}\, U^C\, \partial^{A}_{\cdot} \lambda . \end{aligned}$$

An obvious combination of this result with those of Lemma 3.2 leads to (C 3.3).

Lemma 3.4 *We have*

$$\begin{aligned} U^A \bar{\nabla}_B \partial_A \lambda = \bar{\nabla}_B (U^A \partial_A \lambda) + U^C \partial_C \lambda \, \partial_B \lambda \\ - \exp(-2\lambda) \partial_C (\psi + \lambda) \partial^C \lambda \; U_B . \end{aligned} \tag{C 3.4}$$

Proof. The result is an immediate consequence of the identity

$$\bar{\nabla}_B (U^A \partial_A \lambda) = U^A \bar{\nabla}_B \partial_A \lambda + \partial_A \lambda \bar{\nabla}_B U^A$$

and Lemma 3.3.

Lemma 3.5 *The quantities* $\{U^A K_{AB}\}$ *and* $\{U^A \bar{K}_{AB}\}$ *are related as follows:*

$$\begin{aligned} U^A (K_{AB} - \bar{K}_{AB}) = & -(n-2) \bar{\nabla}_B (U^A \partial_A \lambda) - (n-2) U^A \partial_A \lambda \, \partial_B \lambda \\ & + \exp(-2\lambda) U_B \{(n-2) \partial_A \psi \, \partial^A \lambda - \bar{\nabla}_A \partial^A \lambda\} . \end{aligned} \tag{C 3.5}$$

Proof. The result is obtained by multiplying (C 2.7) by $\{U^A\}$ and then using the previous Lemmas to simplify the result.

4. The Ricci Tensor of $\mathscr{C}_n$

We are now in a position to prove a result of basic importance.

Theorem 4.1 *Let* $\mathscr{H}_n$ *be an n-dimensional, hyperbolic-normal metric space that admits a time-oriented irrotational isometry with basic vector* $\{U^A\}$ *and basic scalar* ψ, *and let* $\mathscr{C}_n$ *be an n-dimensional, hyperbolic-normal metric space that is conformally related to* $\mathscr{H}_n$ *with parameter* λ; *then*

$$\begin{aligned} U^A \bar{K}_{AB} = (n-2)(\bar{\nabla}_B + \partial_B \lambda)(U^A \partial_A \lambda) \\ + \exp(-2\lambda) U_B \bar{\nabla}_A \partial^A (\psi + \lambda) . \end{aligned} \tag{C 4.1}$$

Proof. By Lemma 1.1, the quantity $\{U^A K_{AB}\}$ can be written in terms of $\{U_A\}$ and $h^{AB} \nabla_A \nabla_B \psi$. Combining the results thus obtained with those of Lemmas 3.1 and 3.5, we obtain (C 4.1).

As an immediate consequence of Theorem 4.1, we obtain an evaluation of the Ricci tensor of $\mathscr{C}_n$.

Theorem 4.2 *Under the hypotheses of Theorem 4.1, we have*

$$\bar{K}_{AB} = Z_A U_B + Z_B U_A + (\beta - Z_C U^C) U_A U_B + Y_{AB} , \tag{C 4.2}$$

where

$$\begin{aligned} Z_A &= (n-2)(\bar{\nabla}_A + \partial_A \lambda)(U^B \partial_B \lambda) , \\ \beta &= \exp(-2\lambda) \bar{\nabla}_A \partial^A (\psi + \lambda) , \end{aligned} \tag{C 4.3}$$

and $\{Y_{AB}\}$ *is a symmetric tensor that admits* $\{U^A\}$ *as a null vector.*

Proof. Under the substitution (C 4.3), the system (C 4.1) becomes

$$U^A \bar{K}_{AB} = Z_B + \beta U_B . \tag{C 4.4}$$

Since $\{\bar{K}_{AB}\}$ is symmetric in the indices (A, B), we must have

$$\bar{K}_{AB} = a(Z_A U_B + Z_B U_A) + b U_A U_B + Y_{AB}, \qquad \text{(C 4.5)}$$

for some a and b and for some symmetric tensor $\{Y_{AB}\}$. If we substitute (C 4.5) into (C 4.4), we readily find that $a = 1$, $b = \beta - Z_A U^A$, and that $\{Y_{AB}\}$ must admit $\{U_A\}$ as a null vector.

The following equivalent statement of Theorem 4.1 will be required in the next Section.

Theorem 4.3 *Under the hypotheses of Theorem 4.1, we have*

$$\begin{aligned} U^A \bar{K}_{AB} &= (n-2)(\bar{V}_B + \partial_B \omega - \partial_B \psi)(U^A \partial_A \omega) \\ &\quad + \exp(2\psi - 2\omega) U_B \bar{V}_A \partial^A_\cdot \omega, \end{aligned} \qquad \text{(C 4.6)}$$

where the quantity ω, *defined by the relation*

$$\lambda = -\psi + \omega \qquad \text{(C 4.7)}$$

is termed the λ*-differentia.*

Proof. The system (C 4.6) follows directly from (C 4.1) under the substitution (C 4.7) and use of Lemma 1.1 to evaluate the quantity $U^A \partial_A \psi$.

Theorem 4.4 *Under the hypotheses of Theorem 4.1, necessary and sufficient conditions for* $\{U^A\}$ *to be eigenvector of* $\{\bar{K}_{AB}\}$ *are*

$$\{\bar{V}_B + \partial_B \omega - \partial_B \psi - U_B U^A (\bar{V}_A + \partial_A \omega)\}(U^C \partial_C \omega) = 0. \qquad \text{(C 4.8)}$$

If these conditions are satisfied, the corresponding eigenvalue ρ *is given by*

$$\rho = (n-2) U^A (\bar{V}_A + \partial_A \omega)(U^B \partial_B \omega) + \exp(2\psi - 2\omega)(\bar{V}_A \partial^A_\cdot \omega). \qquad \text{(C 4.9)}$$

Proof. In terms of the λ-differentia, the vector $\{Z_A\}$ defined by (C 4.3) takes the form

$$Z_A = (n-2)(\bar{V}_A + \partial_A \omega - \partial_A \psi)(U^B \partial_B \omega).$$

A decomposition of $\{Z_A\}$ by projection parallel and orthogonal to $\{U_A\}$, together with (C 1.3), (C 4.4) and (C 4.7) establishes the result.

5. The Condition $U^A \partial_A \omega = 0$

The λ-differentia, as defined by (C 4.7), measures the extent to which the conformal relating parameter λ differs from the negative of the basic scalar ψ of the time-oriented irrotational isometry of $\mathscr{H}_n$[5]. The

[5] The reason for comparing λ with $-\psi$ rather than with ψ is due to the fact that the modulus of the vector field that generates the time-oriented isometry in $\mathscr{H}_n$ is taken as $\exp(-\psi)$ rather $\exp(\psi)$.

quantity ω thus determine the extent to which the extent to which the conformal relating parameter can be used to annihilate the basic scalar ψ when considerations are shifted to the space $\mathscr{C}_n$.

Particularly important in applications is the case where the λ-differentia is constant on the trajectories of the irrotational isometry in $\mathscr{H}_n$:

$$U^A \partial_A \omega = 0 . \tag{C 5.1}$$

An obvious sufficient condition for satisfaction of (C 5.1) is

$$\omega = f(\psi) . \tag{C 5.2}$$

Theorem 5.1 *If the λ-differentia satisfies condition* (C 5.1) *and if the hypotheses of Theorem 4.1 are satisfied, then*

$$V^A \bar{K}_{AB} = V_B^{\cdot} \bar{V}_A \partial^A_{\cdot} \omega , \tag{C 5.3}$$

$$V^B \bar{V}_B V_A^{\cdot} = \partial_A \omega , \tag{C 5.4}$$

where the vector field $\{V^A\}$ is defined in terms of the vector field $\{U^A\}$ by

$$V^A = \exp(\omega - \psi) U^A . \tag{C 5.5}$$

We accordingly have

$$V^A V_A^{\cdot} = 1 . \tag{C 5.6}$$

Proof. Under the present hypotheses, Theorem 4.3 gives

$$U^A \bar{K}_{AB} = \exp(2\psi - 2\omega) U_B \bar{V}_A \partial^A_{\cdot} \omega .$$

The system (C 5.3) then follows when $\{U^A\}$ is eliminated by means of (C 5.5) and use is made of (C 3.2) and (C 4.7). Under (C 4.7) and (C 5.5), we have

$$V^B \bar{V}_B V^A = \exp(\lambda) U^B \bar{V}_B (\exp(\lambda) U^A) = \exp(2\lambda) U^B (\bar{V}_B U^A + U^A \partial_B \lambda) .$$

Since (C 1.3), (C 4.7) and (C 5.1) imply $U^A \partial_A \lambda = 0$, a direct application of Lemma 3.3 leads to (C 5.4).

We can now give a very simple geometric characterization of the case in which the λ-differentia is a constant.

Theorem 5.2 *Let the λ-differentia satisfy the condition* (C 5.1) *and let the hypotheses of Theorem 4.1 hold; then the λ-differentia is a constant if and only if the timelike unit eigenvector field of $\{\bar{K}_{AB}\}$ in $\mathscr{C}_n$ generates a geodesic congruence. In this case, the associated eigenvalue of $\{\bar{K}_{AB}\}$ is zero.*
Proof. Under the present hypotheses, Theorem 5.1 shows that the timelike eigenvector field of $\{\bar{K}_{AB}\}$ satisfies the system of equations (C 5.4). Thus, if $\{V^A\}$ generates a geodesic congruence in $\mathscr{C}_n$, we have $\partial_A \omega = 0$,

and conversely. The remaining statement of this theorem is a trivial consequence of (C 5.3).

A straightforward application of previous results leads to a characterization of the λ-differentia in the general case.

Theorem 5.3 *Let the λ-differentia satisfy the conditions* (C 5.1) *and let the hypotheses of Theorem 4.1 be satisfied; then the space $\mathscr{C}_n$ admits a time-oriented irrotational isometry with generating vector*

$$\bar{Y}^A = \exp(-\omega) V^A .$$

Appendix

D. Analysis of the Complete Nonlinear Discretization Equation

1. Statement of the Problem

In view of the scaling and normalization given in Section 32, the complete nonlinear discretization equation assumes the equivalent form

$$\Delta(\varphi) - \Delta_1(\varphi, \varphi) + C\varphi \exp(2\omega - 2\varphi) = 0 \tag{D 1.1}$$

on the surface S_ε of the unit oblate spheroid in Euclidean three-dimensional space. Here $\Delta(\varphi)$ and $\Delta_1(\varphi, \varphi)$ denote the Laplace-Beltrami operators on S_ε, and the constant C is defined by

$$C = \xi r^2, \quad \xi = \frac{2\mu}{\Theta} \exp(-\Pi\,\Theta/\mu). \tag{D 1.2}$$

The problem to be examined in this Appendix is as follows: *Determine single-valued C^2 functions φ on S_ε that satisfy* (D 1.1), *and determine the values of the constant C that lead to such solutions.*

It was shown in Section 32 that S_ε is a compact, oriented, Riemann manifold without boundary and with fundamental metric differential form $d\sigma^2 = m_{\alpha\beta} du^\alpha du^\beta$. We may thus use Bochner's[6] extension of Green's theorem on the manifold S_ε. Accordingly, we have

$$\int \Delta(\varphi) dv = 0 \qquad \int \{\eta \Delta(\varphi) - \varphi \Delta(\eta)\} dv = 0 \tag{D 1.3}$$

for any C^2 functions φ, η, where $dv = \sqrt{\det(m_{\alpha\beta})} du^1 du^2 = dS_\varepsilon$ and the integrations are extended throughout S_ε. These equations provide the basic tools for this investigation.

---

[6] Bochner, S.: *Duke Math. J. 3*, 334 (1937).

2. Conditions for Nontrivial Solutions

Let us make the substitution

$$p = \exp(-\varphi), \qquad 0 < p < \infty . \tag{D 2.1}$$

Eq. (D 1.1) then becomes

$$\Delta(p) = -Cp^3 \ln(p) \exp(2\omega) . \tag{D 2.2}$$

A direct application of (D 1.3) thus gives

$$\int p^3 \ln(p) \exp(2\omega) \mathrm{d}v = 0 . \tag{D 2.3}$$

If we apply Δ to p^2 and use (D 2.2), we have

$$\Delta(p^2) = 2\Delta_1(p,p) - 2Cp^4 \ln(p) \exp(2\omega) .$$

Again, applying (D 1.3), we obtain

$$\int \Delta_1(p,p) \mathrm{d}v = C \int p^4 \ln(p) \exp(2\omega) \mathrm{d}v . \tag{D 2.4}$$

In the case of trivial solutions, there are two possibilities:
(i) $p =$ constant $\neq 1$, $C = 0$,
(ii) $p = 1$, the value of C is left arbitrary.
The case $p = 1$ corresponds to most previous analyses, since (D 1.2) shows that r is undetermined. The situation $p =$ constant $\neq 1$ is easily seen to annihilate the jump discontinuity (i.e., $\mu = 0$ since $r \neq 0$). Noting that a continuous C, and hence a continuous r-spectrum is possible, we henceforth confine our attention to the nontrivial case.

For $p \neq$ constant, the C^2 continuity of p implies that $\int \Delta_1(p,p) \mathrm{d}v$ is a strictly positive functional. It thus follows from (D 2.4) that the value of C is given by

$$C = \int \Delta_1(p,p) \mathrm{d}v / \int p^4 \ln(p) \exp(2\omega) \mathrm{d}v \tag{D 2.5}$$

and that two cases can arise:
(i) $C < 0$, $\int p^4 \ln(p) \exp(2\omega) \mathrm{d}v < 0$,
(ii) $C > 0$, $\int p^4 \ln(p) \exp(2\omega) \mathrm{d}v > 0$.

The determination of the sign of $\int p^4 \ln(p) \exp(2\omega) \mathrm{d}v$ is based on the following easily established identity:

$$\begin{aligned} \Delta(p^k \ln(p)) = {} & p^{k-1}(1 + k \ln(p)) \Delta(p) \\ & + p^{k-2}(2k-1 + k(k-1) \ln(p)) \Delta_1(p,p) . \end{aligned} \tag{D 2.6}$$

Substituting from (D 2.2) into (D 2.6), and using (D 1.3), we obtain

$$\begin{aligned} & \int p^{k-2}(2k-1 + k(k-1) \ln(p)) \Delta_1(p,p) \mathrm{d}v \\ & \quad = C \int p^{k+2}(1 + k \ln(p)) \ln(p) \exp(2\omega) \mathrm{d}v . \end{aligned} \tag{D 2.7}$$

For $k = 1$, this gives

$$\int p^{-1} \Delta_1(p,p)\mathrm{d}v = C \int p^3 \ln(p) \exp(2\omega)\mathrm{d}v + C \int p^3 (\ln(p))^2 \exp(2\omega)\mathrm{d}v\,,$$

and consequently, by use of (D 2.4), we have

$$\int p^{-1} \Delta_1(p,p)\mathrm{d}v = C \int p^3 (\ln(p))^2 \exp(2\omega)\mathrm{d}v\,. \tag{D 2.8}$$

Thus, since p and $\exp(2\omega)$ are necessarily strictly positive, the condition expressed by (D 2.8) can be satisfied only if $C > 0$. We have thus established the following result: *Nontrivial, single-valued, C^2 solutions of* (D 2.2) *exist on S_ε only if $C > 0$, in which case we have*

$$\int p^4 \ln(p) \exp(2\omega)\mathrm{d}v > 0\,. \tag{D 2.9}$$

A further result can be obtained from (D 2.7). For $k = 0$, we have

$$-\int p^{-2} \Delta_1(p,p)\mathrm{d}v = C \int p^2 \ln(p) \exp(2\omega)\mathrm{d}v\,.$$

Hence, since $C > 0$ and $p > 0$, we must have

$$\int p^2 \ln(p) \exp(2\omega)\mathrm{d}v < 0\,. \tag{D 2.10}$$

It is then readily established from (D 2.2), (D 2.3), (D 2.9) and (D 2.10) that p *is a convex function about the trivial solution* $p_0 = 1$; that is, $p-1$ is a convex function on S_ε.

Returning to (D 1.1), and applying the results stated in (D 2.5) we obtain

$$C = \int \Delta_1(\varphi,\varphi)\mathrm{d}v / \int \varphi \exp(2\omega - 2\varphi)\mathrm{d}v\,. \tag{D 2.11}$$

Hence, since $C > 0$, we obtain the inequality

$$\int \varphi \exp(2\omega - 2\varphi)\mathrm{d}v > 0 \tag{D 2.12}$$

as a condition for nontrivial solutions. This, however, is nothing more than the inequality (D 2.10) under the inverse of the substitution (D 2.1). Similarly, from (D 2.3), (D 2.5) and (D 2.9), we obtain

$$\int \varphi \exp(2\omega - 3\varphi)\mathrm{d}v = 0\,, \quad \int \varphi \exp(2\omega - 4\varphi)\mathrm{d}v < 0\,, \tag{D 2.13}$$

and

$$C = -\int \exp(-2\varphi)\Delta_1(\varphi,\varphi)\mathrm{d}v / \int \varphi \exp(2\omega - 4\varphi)\mathrm{d}v\,. \tag{D 2.14}$$

Consequently, a combination of (D 2.11) and (D 2.14) shows that for any non-trivial φ satisfying (D 1.1), we have

$$\int \varphi \exp(2\omega - 4\varphi)\mathrm{d}v \int \Delta_1(\varphi,\varphi)\mathrm{d}v + \int \varphi \exp(2\omega - 2\varphi)\mathrm{d}v \int \exp(-2\varphi)\Delta_1(\varphi,\varphi)\mathrm{d}v = 0\,. \tag{D 2.15}$$

It is thus evident that a nontrivial φ must be a convex function about the trivial solution $\varphi_0 = 0$.

Another representation for C that will be of use in the following analysis is obtained by applying Δ to φ^2 and using (D 1.3). The result is

$$C = \int (1 + \varphi) \Delta_1 (\varphi, \varphi) \, dv / \int \varphi^2 \exp(2\omega - 2\varphi) \, dv, \qquad \text{(D 2.16)}$$

and it has the property that the denominator is a strictly positive functional.

3. Expansion in the L_2-Norm

We now come to grips with the problem of existence and uniqueness of nontrivial solutions of (D 1.1). The basic question is easily seen to be that of the bifucation of nontrivial solutions from the trivial solution ($\varphi = 0$, $C =$ all real numbers). In the case where we look for solutions that are independent of the elliptic longitude, the results of Keller[7] can be used. In the general case, we can use the results of Berger[8]. With minor modification for the fact that S_ε is compact and without boundary, either method shows that nontrivial solutions of (D 1.1) exist, provided their norm (Sobolev or L_2[9]) is sufficiently small, and that these nontrivial solutions bifucate from the trivial solution only at the *eigenvalues of the linear approximation of* (D 1.1).

In view of the nonlinearity of (D 1.1), the solutions of this equation are intrinsically dependent on the specification of the norm of φ (see previous footnote), while the only information available about φ is that φ must be a single-valued C^2 function on S_ε. In terms of the results established by Berger, it is clear that any expansion procedure should be in terms of the appropriate norm as the parameter. We shall follow this course using the L_2-norm as the expansion parameter[10]. If we make the substitution

$$p = 1 + \hat{\rho} \qquad \text{(D 3.1)}$$

[7] Keller, J. B.: Bifucation Theory of Ordinary Differential Equations, in: *Bifucation Theory and Nonlinear Eigenvalue Problems* (ed. Keller, J. B., and Autman, S.), New York 1969.

[8] Berger, M. S.: A Bifucation Theory for Nonlinear Elliptic Partial Differential Equations and Related Systems, in: *Bifucation Theory and Nonlinear Eigenvalue Problems, op. cit.*

[9] An integration of (D 1.1) over S_ε and use of Bochner's theorem gives

$$\int \Delta_1 (\varphi, \varphi) dv = C \int \varphi \exp(2\omega - 2\varphi) d\,v,$$

and hence the L_2-norm of the "first derivatives" is controlled by the L_2-norm of φ.

[10] This is the course followed in Appendix B of Edelen, D. G. B.: Relativistic Galactic Morphology I: General Theory, RM-4017-RC, The RAND Corporation, 1964. This procedure includes the cases where there is an expansion in the amplitude of the solution of the associated linearized equation; see, Millman, M.: Pertubation Solutions of Some Nonlinear Boundary Value Problems. *Bifucation Theory*

in (D 2.2), we have

$$\Delta(\hat{\rho}) = -C(1+\hat{\rho})^3 \ln(1+\hat{\rho}) \exp(2\omega) \tag{D 3.2}$$

and hence the trivial solution ($\hat{\rho} = 0$, $C =$ all reals) corresponds with the trivial solution ($\varphi = 0$, $C =$ all reals) of (D 1.1). We shall base our considerations on (D 3.2) rather than (D 1.1) since the analysis based on (D 3.2) is simpler[11].

Let $l(\eta)$ denote the L_2-norm of an L_2 function η on S_ε, that is $l(\eta)^2 = \int \eta^2 \, dv$, and set

$$L = l(\hat{\rho} \exp(\omega))\,^{12}. \tag{D 3.3}$$

Under the substitution

$$\hat{\rho} = LU, \tag{D 3.4}$$

we may replace $\hat{\rho}$ by a function U with given norm

$$l(U \exp(\omega)) = 1, \tag{D 3.5}$$

and thereby consider L as a parameter. From (D 3.2) and (D 3.4) we must then have

$$L\Delta(U) = -C(1+LU)^3 \ln(1+LU) \exp(2\omega), \tag{D 3.6}$$

with U a single-valued, C^2 function on S_ε such that (D 3.5) holds. Since p must be positive, it is understood that L is to be chosen such that $L \min_{S_\varepsilon}(U) > -1$.

In general, U and C are functions of L. For the remainder of this study, we examine those solutions of (D 3.6) that can be expanded in a formal power series in L for L in some interval $(0,b)$ with b strictly positive. That this precedure is valid for sufficiently small L follows from Berger's cited results. We thus have

$$U = \sum_{i=0}^{\infty} U_i L^i, \quad C = \sum_{i=0}^{\infty} C_i L^i \tag{D 3.7}$$

with

$$k!\,U_k = \partial^k U/\partial L^k|_{L=0}, \quad k!\,C_k = \partial^k C/\partial L^k|_{L=0}. \tag{D 3.8}$$

---

and *Nonlinear Eigenvalue Problems, op. cit.*; Millman, M. H., Keller, J. B.: *J. Math. Phys.*, January 1969; Keller, J. B., Ting, L.: *Comm. Pure Appl. Math.* 19 (1969). It should be noted that there are cases where the expansion in terms of the amplitude of the solution of the associated linearized equation does not provide the correct results while the expansion in terms of the L_2-norm of the solution works without difficulty.

[11] Although the series representation of the eigenvalues is the same for (D 3.2) and (D 1.1), the series for the solutions converges more rapidly when (D 3.2) is used. Further, since (D 3.2) is variational, Berger's (*op. cit.*) results show that a bifucation always exists at each eigenvalue of the linearized equation.

[12] The inclusion of the function $\exp(\omega)$ results in significent simplification. Since ω and $\hat{\rho}$ are of class C^2 on S_ε, we clearly have that $(\hat{\rho}\exp(\omega))$ belongs to L_2.

For obvious reasons, we refer to the series (D 3.7) as expansions in the L_2-norm.

If we substitute (D 3.7) into (D 3.5), the result is

$$1 = \int \sum_{j=0}^{\infty} \left(\sum_{i=0}^{j} U_i U_{j-i} \exp(2\omega) \right) L^j \mathrm{d}v .$$

Since this must hold for all L in $(0,b)$, we must have

$$\int U_0^2 \exp(2\omega)\mathrm{d}v = l(U_0 \exp(\omega)) = 1 \tag{D 3.9}$$

and

$$\int \sum_{i=0}^{j} U_i U_{j-i} \exp(2\omega)\mathrm{d}v \equiv J_j = 0 \quad \text{for} \quad j \neq 0 .$$

These relations provide the norms and the "orthogonality" relations among the elements of the sequence $\{U_i\}$.

4. The First Approximation

For $i = 0$, (D 3.6), (D 3.7) and (D 3.8) give

$$\Delta(U_0) = -C_0 U_0 \exp(2\omega) , \tag{D 4.1}$$

and consequently (D 1.3) yields

$$\int U_0 \exp(2\omega)\mathrm{d}v = 0 . \tag{D 4.2}$$

If we use (D 4.1) to evaluate $\Delta(U_0^2)$ and again use (D 1.3), the result is

$$C_0 = \int \Delta_1(U_0, U_0)\mathrm{d}v , \tag{D 4.3}$$

since (D 3.9) yields

$$\int U_0^2 \exp(2\omega)\mathrm{d}v = 1 . \tag{D 4.4}$$

In the case where S_ε is the unit sphere, S_0, and $\exp(2\omega) =$ constant from spherical symmetry, the only single-valued, C^2 solutions of (D 4.1) on S_0 are the spherical harmonics $P_n^m(\cos u^1)\exp(imu^2)$, with m and n integers and $0 \leq |m| \leq n$, these solutions obtain only if C_0 assumes the discrete values $\{n(n+1)\exp(-2\omega)\}$, and the n-th eigenvalue is of multiplicity $2n+1$. In the general case, we have

$$C_0 = \rho(n,m,\varepsilon)^2 , \quad 0 \leq |m| \leq n , \quad \rho(n,0,0)^2 = n(n+1) \tag{D 4.5}$$

and the eigenvalues are simple whenever $\varepsilon > 0$.

5. The Second and Higher Order Approximations

For $i = 1$, (D 3.6), (D 3.7) and (D 3.8) give

$$\Delta(U_1) + C_0 \mathrm{e}^{2\omega} U_1 = -C_1 \mathrm{e}^{2\omega} U_0 - \tfrac{5}{2} C_0 \mathrm{e}^{2\omega} U_0^2 . \tag{D 5.1}$$

Integrating this equation over S_ε and using (D 1.3), (D 4.2) and (D 4.4), we obtain

$$\int e^{2\omega} U_1 \, \mathrm{d}v = -\tfrac{5}{2}, \tag{D 5.2}$$

while (D 3.10) yields

$$\int e^{2\omega} U_0 U_1 \, \mathrm{d}v = 0. \tag{D 5.3}$$

Both the solvability condition for (D 5.1) and the condition (D 5.3) are effected when (D 5.1) is multiplied by U_0, integrated over S_ε, and use is made of (D 1.3). The result gives the following determination of C_1:

$$C_1 = -\tfrac{5}{2} C_0 \int e^{2\omega} U_0^3 \, \mathrm{d}v. \tag{D 5.4}$$

When the same thing is done starting with a multiplication by U_1, we then obtain

$$C_0 \int e^{2\omega} U_1^2 \, \mathrm{d}v = \int \Delta_1(U_1, U_1) \, \mathrm{d}v - \tfrac{5}{2} \int e^{2\omega} U_0^2 U_1, \tag{D 5.5}$$

and hence we always have

$$\int \Delta_1(U_1, U_1) \, \mathrm{d}v > \tfrac{5}{2} \int e^{2\omega} U_0^2 U_1 \, \mathrm{d}v.$$

Proceeding in a similar fashion, we easily establish the following results:

$$\begin{aligned} 6(\Delta(U_2) + C_0 e^{2\omega} U_2) = & -6 C_2 e^{2\omega} U_0 - 15 C_1 e^{2\omega} U_0^2 \\ & - 11 C_0 e^{2\omega} U_0^3 - 6 C_1 e^{2\omega} U_1 - 30 C_0 e^{2\omega} U_0 U_1, \end{aligned} \tag{D 5.6}$$

$$\int e^{2\omega} U_2 \, \mathrm{d}v = \frac{11}{15} \frac{C_1}{C_0}, \tag{D 5.7}$$

$$\int e^{2\omega} U_0 U_2 \, \mathrm{d}v = - \int e^{2\omega} U_1^2, \tag{D 5.8}$$

$$C_2 = C_0 \left\{ \tfrac{75}{12} \left(\int e^{2\omega} U_0^3 \, \mathrm{d}v \right)^2 - \int e^{2\omega} U_0^2 \left(\tfrac{11}{6} U_0^2 + 5 U_1 \right) \mathrm{d}v \right\}, \tag{D 5.9}$$

where (D 5.9) obtains as a consequence of the solvability condition for (D 5.6), it being noted that U_0 is a solution of the corresponding homogeneous equation.

When the various results obtained above are combined, we obtain the following representation of the eigenvalues:

$$\begin{aligned} C = C_0 \Big[1 & - \tfrac{5}{2} \int e^{2\omega} U_0^3 \, \mathrm{d}v \, L \\ & + \left\{ \tfrac{75}{12} \left(\int e^{2\omega} U_0^3 \, \mathrm{d}v \right)^2 - \int e^{2\omega} U_0^2 \left(\tfrac{11}{6} U_0^2 + 5 U_1 \right) \mathrm{d}v \right\} L^2 \Big] + 0(L^3). \end{aligned} \tag{D 5.10}$$

From this we see that for L sufficiently small, which, within the theory of galactic morphology given in Chapter III is equivalent to the statement that there is small deviation from oblate spheroidal symmetry, the eigenvalues of the reduced discretization equation are essentially those of the linear approximation. Since the same results hold for the associated eigenfunctions, we are permitted to base our considerations on the linearized version of the reduced discretization equation when describing observed objects as small deviation from oblate spheroidal symmetry.

Author Index

Subject Index

Springer Tracts in Natural Philosophy

Vol. 1 **Gundersen: Linearized Analysis of One-Dimensional Magnetohydrodynamic Flows**
With 10 figures. X, 119 pages. 1964. Cloth DM 22,– ; US $ 6.10

Vol. 2 **Walter: Differential- und Integral-Ungleichungen** und ihre Anwendung bei Abschätzungs- und Eindeutigkeitsproblemen
Mit 18 Abbildungen. XIV, 269 Seiten. 1964. Gebunden DM 59,– ; US $ 16.30

Vol. 3 **Gaier: Konstruktive Methoden der konformen Abbildung**
Mit 20 Abbildungen und 28 Tabellen. XIV, 294 Seiten. 1964
Gebunden DM 68,– ; US $ 18.70

Vol. 4 **Meinardus: Approximation von Funktionen und ihre numerische Behandlung**
Mit 21 Abbildungen. VIII, 180 Seiten. 1964. Gebunden DM 49,– ; US $ 13.50

Vol. 5 **Coleman, Markovitz, Noll: Viscometric Flows of Non-Newtonian Fluids.**
Theory and Experiment
With 37 figures. XII, 130 pages. 1966. Cloth DM 22,– ; US $ 6.10

Vol. 6 **Eckhaus: Studies in Non-Linear Stability Theory**
With 12 figures. VIII, 117 pages. 1965. Cloth DM 22,– ;US $ 6.10

Vol. 7 **Leimanis: The General Problem of the Motion of Coupled Rigid Bodies About a Fixed Point**
With 66 figures. XVI, 337 pages. 1965. Cloth DM 48,–; US $ 13.20

Vol. 8 **Roseau: Vibrations non linéaires et théorie de la stabilité**
Avec 7 figures. XII, 254 pages. 1966. Reliée DM 39,– ; US $ 10.80

Vol. 9 **Brown: Magnetoelastic Interactions**
With 13 figures. VIII, 155 pages. 1966. Cloth DM 38,– ; US $ 10.00

Vol. 10 **Bunge: Foundations of Physics**
With 5 figures. XII, 311 pages. 1967. Cloth DM 56,– ; US $ 14.80

Vol. 11 **Lavrentiev: Some Improperly Posed Problems of Mathematical Physics**
With 1 figure. VIII, 72 pages. 1967. Cloth DM 20,– ; US $ 5.50

Vol. 12 **Kronmüller: Nachwirkung in Ferromagnetika**
Mit 92 Abbildungen. XIV, 329 Seiten. 1968. Gebunden DM 62,– ; US $ 17.10

Vol. 13 **Meinardus: Approximation of Functions: Theory and Numerical Methods**
With 21 figures. VIII, 198 pages. 1967. Cloth DM 54,– ; US $ 14.80

Vol. 14 **Bell: The Physics of Large Deformation of Crystalline Solids**
With 166 figures. X, 253 pages. 1968. Cloth DM 48,– ; US $ 12.00

Vol. 15 **Buchholz: The Confluent Hypergeometric Function** with Special Emphasis on its Applications
XVIII, 238 pages. 1969. Cloth DM 64,– ; US $ 16.00

Vol. 16 **Slepian: Mathematical Foundations of Network Analysis**
XI, 195 pages. 1968. Cloth DM 44,– ; US $ 12.10

Vol. 17 **Gavalas: Nonlinear Differential Equations of Chemically Reacting Systems**
With 10 figures. IX, 107 pages. 1968. Cloth DM 34,– ; US $ 9.40

Vol. 18 **Marti: Introduction to the Theory of Bases**
XII, 150 pages. 1969. Cloth DM 32,– ; US $ 8.80

Vol. 20 **Edelen, Wilson: Relativity and the Question of Discretization in Astronomy**
With 34 figures. XII, 186 pages. 1970. Cloth DM 38,– ; US $ 10.50